I0464350

Oscillators-in-a-Substance Model of Existence
A Physicists' Grail and An Alexander's Sword

Dean LeRoy Sinclair (BA, MS, PhD)

A viewpoint change on century-old data reveals a comprehensive model of existence.

Volume 1 of "Eski Eshek Konusuyor" series (EEK-1)
Edition One, Large Print

...

PREFACE

It is probable that when physicists think of the finding of a "Physicist's Grail,"
they think that what will be found is some set of Differential Equations which, as a set of mathematical equations, will contain the wisdom of some great genius, unifying the "Four Forces of Nature" and otherwise satisfying the expectations that seem to have developed for a scientific theory.

Unfortunately, the reality is different. It turns out that all that is needed are some simple changes in viewpoint which should have been obvious in the early years of the 20th Century, rather than appearing here in the middle of the second decade of the 21st Century,
The resulting "Explanation of Almost Everything."
called herein "The Oscillators-in-a-Substance Model" uses only the simplest of basic Mathematics and is so sweeping in its implications--changing views which have been accepted for many years-- that one contributor to this work, in a private letter, called it ,
"an Alexander's Sword" for cutting through the Gordian Knot of Tangled Theories that are confining the Physical Sciences

 Hence the "Working Title"given to this book, The

Oscillators-in-a-Substance Model   A Physicist's Grail /An Alexander's Sword.

It is hoped that this little book will be found to be interesting reading and valuable as a source of ideas for discussion. Undoubtedly, parts will later be found to need revision. Some ideas and concepts that were accepted by this writer at the start, in the Spring of 2004, were later modified, or discarded, as information and insights arose.

Welcome to the adventure.

PRELUDE

if you understand completely what is going on in the following essay, there is a possibility that you do not really need to read the rest of the book. However, it is hoped that you will find it to be interesting, anyhow. If the discussion below-- which turns out to be a brief summary of the major points of the book-- has too much technical jargon, and is too condensed, please don't despair of the rest of the book. it tries to be at a more readable level.

A Little Poem with Big Consequences

A careful look at the possible messages hidden in the first verse of an innocent-seeming poem keys into a quite complete "Theory of Everything "

A, poem published on the Internet a few years ago, started with the following verse:

> "On some long-bygone day,
> Someone set up a 3-D, dot array,
> picked a spot,
> plucked a dot;
> and Creation was on its way."

At first glance, the above is simply a bit of Limerick-like doggerel. Then, it was noticed that there was a mono-theistic approach, which combined two, apparently-antithetical positions, "Intelligent Design" and "Evolutionary Creation," with Chaos Theory, in only five lines.
The thought occurred that this little verse might contain far more depth than one would ever be expected to notice. What would happen if we assumed it to be essentially true, keyed into Our Universe?
  What would be the necessary characteristics of the "3-D dot array?" What must happen with the "plucked dot, " for Creation to continue from one plucking?

What is the implication of the choice of the word, "Pluck?"

What if our existence be within a 3-D array? The Void of Space, in fact, Everything, would be full of this "Something." rather than "Empty where Matter does not exist," as is currently assumed by scientists. In fact, Matter would have to consist of some sort of clumping of dots.

The 3-D array would have to be capable of carrying Information/Energy by electromagnetic radiation, at the "Speed of LIght." An array that did this could be possible if the "dots" were rotors in contact with one another which were rotating at an average tangential velocity of the Speed of LIght.

As Light is considered a transverse wave, our array has to have a characteristic of a solid, However, it seems to have considerable flexibility as "Matter" objects move within it.   This sounds much like a chemical substance at its Triple Point where the substance, while usually in a liquid state, may act as a solid or gas depending on variation of pressure. LIght then would be a pressure disturbance which creates its own Solid/Gas Wavefront as it moves.

We know that "The only unchanging thing is Change, .For there to be Existence, there must be continuous change. "

What about the "Plucked Dot?" The term," pluck," has an implication of "to start vibrating." That Plucked Dot would need to continue to vibrate, furnishing the "motive power," the work, to keep Creation going. It could be called "The Controlling Oscillator of our Universe."

This Control Oscillator idea can account for a number of phenomena including the mysteries of "Black Matter" and "Dark Energy." The location of the "Plucked-Dot-Oscillator" would be in the center of the volume which is known to furnish most of the Microwave Energy Background of Our Universe. That would be the volume in which most of the matter of our Universe was still concentrated, as the postulated Control Oscillator not only furnishes motion to the matrix, but, also, rearranges parts of the matrix into units, which give rise to neutrons. These, in turn, give rise to the electron and proton units which rejoin to Hydrogen Atoms. These, eventually, coalesce into all of the units that we know as Matter. As Matter created in the region of the Control Oscillator accumulates, the motion of the oscillator will cause it to expand out through the original matrix.

At our position in the matrix, we have observations which we may use to determine things about "Existence," and more about our postulated "Control Oscillator."

Let us go back to the idea of a Substance. Within a substance would be a pressure, A pressure which is always varying slightly but averaging out to a "Force which is its own equal and opposite." This universal One Force, gives us a definition for our term, "Mass," as a measure of the internal pressure within a unit, against the pressure of the Remainder of the Universe (or Matrix) which allows the unit to have existence. This pressure would be felt, to some extent, by every other unit in the matrix, and accounts for what we know as Gravity. Gravity would be the apparent attraction felt by any two entities because there would be less of the matrix on the line between the two units than there would be behind them on the same line. Their pressures against the matrix would tend to force apart the units between creating an apparent attraction. This Gravitational Force of Attraction will disappear when the two units come into physical contact with one another, at which point they operate as one unit.

This Gravitational Force will operate between any two units and will operate within any set of units that may be considered. In a diffuse, but coherent, unit, such as a Galaxy, the "force cones" will balance at the physical center of the unit to create an apparent absence of force-- or presence or of a huge force equivalent to the total mass of the object--either view will give the impression of an attraction of every unit of the Galaxy into a "Hole" or toward a physical mass, neither one of which is there, There is only the summation of vectors, equivalent in size to the entire mass of the Galaxy.

Electromagnetic Radiation seems to be the major motion within our matrix. We can suspect that it originates from our postulated Control Oscillator.

There are two Constants of Nature closely related to Electromagnetism. The Speed of Light, "c," and Planck's Constant, "h." The latter may be taken to be the "angular momentum, per cycle," of an information wave riding on electromagnetic waves traveling at an average velocity of "c." One definition of angular momentum is "Mass times velocity times radius." So we write, "h=mvr/cycle." Since the Speed of Light, "c," is, apparently, the underlying velocity of the carrier wave, we assume it to be the tangential velocity of the rotors, and insert c for v to obtain, h=mcr/cycle, and rearrange to form a new constant, "h/c per cycle" which has the dimensions of mass times radius per cycle. Noting that this is independent of the length of the cycle and has the dimensions of Energy or Work (Mass times Distance) we may define this as a basic Quantum of Existence.

This could be a measure of the work being done, continuously, by the Control Oscillator. It could be called the "Basic Quantum which gave the name to the "Quantum Revolution of the Twentieth Century." Also, in the form of m x r =h/c, the equation could be taken as the definition of a family of oscillators and used to find the limits. Doing this with the electron and the proton, considering Rest Mass to be a possible limit,"m." the corresponding "r" can be seen to be the "Compton Wavelength."

  This gives a connection/explanation of the Particle/Wave Duality of the electron and proton. Considered as coherent units, "mass times radius," they are particles. Considered as oscillators, with a limit wavelength of the same dimension as the radius, they show vibration, hence the wave characteristics. Using the idea, " If one limit can be found , a reciprocal limit may exist at the situation where the coefficients are reversed," one finds that the electron and proton would contain the same basic work function but expressed in very different ways, the electron having a range some 1832 times that of the proton. The electron would be both much larger and much smaller than the proton, and much lighter and much heavier than the proton.   This gives rise to the possibility that the "nuclear atom" consist of what might be called "Two Interpenetrated, Plasma

Droplets,." The tiny center one defined by the clumping of the protons, the outer dimensions defined by the range of vibration of the electrons, which would be, also, vibrating through and into the center if the proton cluster.

A plasma is a cluster of charged particles. The electron and the proton have "Charges."   We need to determine what we are talking about when we use that term "Charge." Since we see two kinds of "Charge," we assume that the charges represent two opposite facets of the units, particularly when each is known to have an apparent, exact, or almost exact , partner unit with the "opposite charge." Opposite units would be mirror images. Left or right-handed images, or of opposite rotational senses, or both. If these units are vortex oscillators, which would arise from the splitting of a single oscillator having counter-rotating opposite halves, then charge would be a demonstration of this. The supposed "Annihilation" of electron and anti-electron-- after first joining into a neutral "molecule--" suggests exactly this. There is first joining into a neutral molecular-- two-center-- form which then combines into a neutral atom, an oscillator having one center. That center, however, may be within an equator and the atom, at any instant, consist of two counter-rotating halves which can be split. This splitting is also known, it is called "pair-production."

We. can go back to the basic quantum, "h/c" and check on the idea that an average value of the radius and mass could possibly define the characteristics of the basic control oscillator. One average value would be found by taking the square root of the value of "h/c." As the value of h/c, in the cgs system, is about 2.2×10^{-37} g. cm. an "average figure" would be about 4.7×10^{-19} g. or cm. This suggests that the control oscillator may have a basic wavelength of about 4.7×10^{-19} cm. corresponding to a frequency of about 6.4×10^{28} cps.

Starting from speculations engendered by a little poem, we appear to have progressed to a simple "Theory of Everything" having wide applicability.

$\div$

PART ONE. THE BASIC IDEAS.

INTRODUCTION

During the last few years, a number of small papers have been published on the Internet that bear on a model currently called the "Oscillators-in-a-Substance Model." This is a "Theory of Everything," which could be called " A Physicists' Grail," or "An

Alexander's Sword for the Tangle of Physics Theory."

Although this model "grew itself," as the result of a "Twilight Years Hobby of Rewriting Physical Science Theory," it may have some interest to others.
This is an attempt to collect the major points of these into a readable, coherent manuscript covering the essentials of that "Theory of Everything ," and its quite wide implications.

As this book is being put together from earlier manuscripts rather than being totally written "from scratch," there will be some repetition and some cases of slightly different viewpoints as the concepts changed slightly as they evolved over the last eleven years (2004-2015.)

The Logic Train Leading to Basic Ideas of the Model

The trail of thought leading to the hobby, and to this manuscript, started in the Spring of 2004 with an idea that Einstein's Genius might have been his ability to see the significance of the "Overlooked Obvious." Although this idea appears to have been in error; the idea of looking at the "Obvious," to see if something else might be there, turned out to be useful.
The first thing that was looked at started a train of

logic that led far beyond what could ever have been anticipated.

The first thought was, "Is there anything in the work of Einstein, himself, that is obvious, with something in it that might be overlooked?"

From there the thoughts went much as follows:
Much of Einstein's Work seems to be based on these equations which include the v^2/c^2 factor.
  Any equation could be generalized.
  What would happen if these equations were to be generalized in some way?

  What then would be the significance of the constant, "c"?
These equations would apply to any case where information was transferred with a maximum velocity of transfer represented by the constant which was applicable to the situation.
  That is, these equations would apply to the distortion of information transferred in water, in air, by nerve impulse, by Pony Express. In fact, they would apply to any "Domain or Perceptual Universe Defined by a Maximum Velocity of Information Transfer."
What is common to all information transfer?
  Information is transferred in "packets" from one

carrier to another,

What is the maximum velocity that information can be transferred over a fairly long distance? That would be the average velocity of the information carriers. That is, for a one-way message.

What would that say about the "Speed of LIght?" The Speed of LIght, in a Vacuum, should be an average velocity of whatever the information carriers are within a "Vacuum."

Electromagnetic Radiation is known to carry in straight lines, spreading out in all directions from a point radiator. What kind of carriers could do this, at the same time as producing what appears to be a transverse wave motion? This is a double question, we have to look at it in two parts.

To have a transverse wave motion, we need something which can act as a solid.

How can a vacuum be a "solid," which would carry such a wave motion?

After considerable speculation, an answer arose to this question.

The above conditions are met by a "Substance, at its Triple Point" where it may act as Liquid, Solid or Gas, depending on the pressure. This would be a substance which is composed of rotors having a rotational (tangential) velocity averaging the Speed of LIght.

This Substance would carry the " Pressure Disturbance Called LIght," as if it were a solid, as a wave motion riding on the tangential velocity contacts of the units. This pressure disturbance would create a varying Solid/Gas Front in the basic Liquid Substance.

From this insight as to the probable nature of Light, and its transmission in a Vacuum has arisen "The Oscillators-in-a-Substance Model of Existence" which redoes much of basic Physical Science Theory.

The assumption of a Basic Substance, of unknown extent and undefined simplest unit, immediately states the limits of the model. It makes no claims to define the Fact of Existence. However, from the idea of a basic substance--plus the fact that , "The only thing that does not change is Change," there can be seen an immediate solution to Einstein's Unsolved Problem of Unification of Forces.

There would be but one True Force-- "Its Own Equal and Opposite--" a very slightly varying Universal Pressure within that Substance.

From this, there appears immediately a definition of "Mass" as an instantaneous measure at a point on a surface of an entity of the pressure within that entity. this is a force against the rest of Existence that allows that particular entity to survive.

This definition of Mass defines "Gravity" very simply, as "The apparent attraction between Matter Entities due to the fact that there is less of the Substance of Existence between any two units that there is behind the two on the line connecting them."
 That is, the pressure of an entity against the rest of existence is felt "radiating out forever" and will be felt, to some extent, "Forever." The force of the masses, as felt between the two units, will tend to push aside the material between them, and they are forced toward each other by the material behind them. This interaction between two units disappears when the units touch, and, thereafter, act as one unit.

The fact of radiation outward of the pressure of the Mass of an Entity when applied to a coherent-but-diffuse unit such as a Galaxy, accounts for the phenomena, which are said to be "Black Holes." The Radiation into the Center of the Forces of Masses, may be considered as if it were made up of cones focused into the center from every unit. The result is, at the Physical Center of Balance, the object has an instantaneous point where all of the mass of the unit acts as if it were concentrated there.

This combined action be a Vector Summation of Masses just outside, and surrounding, the point where they balance. In a Galaxy, as the units of the Galaxy are moving at all times, the instantaneous Center of Mass will sweep out a volume. This Black-Hole Volume, if one wishes to retain the name, is just that, a volume. It is not an infinitely deep pit.
The associated huge, invisible mass observed there is the Vector Mass Summation spoken of above, not a physical entity.

What we have discussed thus far, are results that could easily have been inferred years ago, had the Michelson-Morley Experiment of 1890 been interpreted as giving information as to what was present where Matter was not, rather than by the assumption, "Where there is no Matter, there is Nothing."

It seems as if some reasonably intelligent, Secondary School student should have noticed, at some time during the past Century, that there was a much simpler and more logical pattern to Existence than was being taught in the schools. The data discussed above was available. Also has been available the work of Max Planck, which when slightly reinterpreted, adds the idea of "Oscillatory Motion" to the motion necessary for existence and adds the second "Dimension" to the Oscillators In-a-Substance Model.

It is hoped that this little introduction will have intrigued you enough to go on to see the additional insights and explanations that arise when the work of Max Planck is integrated with the ideas introduced above.

The only paper published in a Scientific Journal on the Oscillators- in-a-Substance Model, was included in the Nov./Dec. 2013 Issue of Infinite Energy, *as Paper Number Nine in a special issue on papers having to do with the Theory of Cold Fusion.*
Unfortunately,the Galley Proof was not edited and that paper has a number of embarrassing typographical errors. A somewhat cleaned up version was published on the Oscillator/Substance Theory Group Site. which was a site set up in 2008 to explore the basic ideas covered here.

That version is reproduced here--in an additionally cleaned up form.

Some Implications of the Oscillators-in-a-Substance Model

The Oscillators in a Substance Model is a theoretical approach which seems to explain everything--except the fact of existence. It is a model which has been "hidden in plain sight," for over a century. It has been obscured by the misinterpretation of the Michelson-Morley experiment as proving that the "void" was empty space.

(This model might well have come into existence over a century ago had it been realized that the disturbances that we call "light" could be being carried by rotors in contact with one another and that the work of Max Planck could be used to further define those rotors. Einstein made a couple of errors of judgment and the scientific community followed his lead.)

A summary statement for this model is, as follows: "All existence is considered as being within a Substance/Substrate of unknown extent and undefined basic unit, which has the general characteristics of a chemical substance at its triple point. This Substance is organized into and/or controlled by oscillators of the family defined by the equation, $m \times r = h/c = r \times m$, with an inversion through the value defined by $(h/c)^{0.5}$

This square root of the ratio of Planck's Constant to the Speed of Light describes a hypothetical entity with *a mass of about 4.7 x10^{-19} grams and a radius of about 4.7 x 10^{-19} cm, This entity would be the average size of the oscillators of our existence, and represents an average mass value.*

This model has many implications in physics theory. One is that what we consider as the rest mass, is apparently a minimal pressure against the rest of existence exerted at the maximal radius of the oscillator. This maximum radius turns out to be in the literature as the "Compton Wavelength," for many units. The probable other limit, of maximum pressure at minimal radius, can be estimated by reversing the absolute values.

This approach leads to the interesting conclusion that the electron has limits such that it will, in a single inversion /rotation/inversion-of-rotation cycle be able to pass not only through and into an atomic nucleus but actually far inside of a proton or neutron. The electron/positron pair appear to be interconvertible. Also,the idea of matter and antimatter being annihilated appears to arise from a situation of an electron/anti-electron pair reaching an exact alignment such as to combine to a previously unsuspected parent unit, which may be the ubiquitous unit of the void. A unit possibly distortable to the neutron by shock wave events. Considered this way, this unit, dubbed by this writer, the "Zerotron." would

be, in effect, the parent unit of all "Matter."
[Experimentalists need to be aware of this probable,
ubiquitous denizen of the "Null Set." in any work.
Especially if their work will in any way cause a shock
wave, resulting in a possible neutron burst. The
radiation sickness reported by leClaire as a result of
one of his experiments with cavitation collapse may
well be an example of being unaware of this factor.
It seems probable that while stars are converting
Hydrogen through a number of steps to Iron, they are
also converting more Zerotrons to neutrons,
furnishing more Hydrogen to burn.
It might be interesting to check to see if such simple
shock-wave producing acts as an electrical discharge
in a vacuum and striking an anvil with a hammer
would produce some neutrons.]

The pressure of the 'Units of the Void" on the
comparatively infinitesimal amount of itself which we
know as "Matter" is the "One Force Which is Its Own
Equal and Opposite." That is there is always a
pressure toward equalizing motion throughout
existence. On a localized level, this pressure is felt as
a push toward a common center of least net motion.
The commonest manifestation of this we call
"Gravity," which is usually considered as an attraction
between matter units. This mathematical description
works well, except at the limit. *At the limit, the*

pushing-together concept reaches a balance. The "pull together" formulation, however, implies an infinitely deep hole.

(This error in the concept of Gravity causes the Black Hole Misconception of the astronomers. There is nothing wrong with their data. They need to switch their view of Gravity from a "pulling together" to a "pushing together" to understand why they cannot find "the huge matter object that 'should be there.'")

The realization that Matter is but an almost infinitesimal part of all that exists, as well as the realization that what we consider as the rest mass of a unit is a tiny fraction of the average mass accounts for a number of "Inexplicable mysteries," e.g., " Black Matter," and "Dark Energy." Energy, in this model, is another name for a Packet of Motion. Mass is the name for the accumulated pressure effects of the motions within a surface as expressed against the rest of existence at a point on that surface. It is usually measured by comparison to some "standard." Energy--usually meaning the motion effects observed when moving units collide--is measured by collision effects.

Charge is defined in this model as the observed manifestation of the residual rotation associated with one of the vortex oscillators forms. A net counter-clockwise residual effect is probably the "negative charge" associated with electrons. A net clockwise rotation would be a positive charge.

It may be noted that while charges are associated with an individual unit, the effects that we see and attribute to the individual units are effects of huge numbers of units together. At the level of individual units or small groups of units the situation is somewhat different. There is evidence that electrons pair, and there is no reason to think protons do not pair, also. The Positronium unit may be essentially the same as the electron pair noted for molecules. except that, on the molecular level, or even the atomic level, interaction with other units prevents the perfect alignment that would allow the ultimate combination called "annihilation."

By this model, it appears that people, with the equipment to do so, could start with electrons and by a progressive acceleration in an electromagnetic field, follow "Steps of Evolution" as some of these electrons changed into "anti-electrons," protons, anti-protons, muons and anti muons, zeta particle etc as the "electron" units were alternately accelerated and compressed ("increased in mass") by interaction with the field which would have a constant rotational aspect.This constant rotational aspect would have, alternately, directional acceleration and compression effects on the electron-antielectron moiety during its rotation/inversion-inversion-of-rotation cycle.

 Tracking should show signals attributable to the anti-electron, the Positronium Unit in its two forms, some "Annihilation radiation," signals for a proton, then the antiproton, their combination to a "positronium analog," etc. The situation becoming more and more complicated as the "acceleration pressure" increases. The seekers for the Higgs Boson, who started at the "proton" level may have come close to producing the "idealized average unit" which would be supposedly reached at an ultimate compression to the inversion value of $(h/c)^{0.5}$.

 (In a sense, as the central,average unit in this model, this central unit has the same effect of "Supplying Gravity," to the Substance of Existence as the Higgs Boson does to the more fanciful Standard Model Set of Units. (The supposed evidence for Quarks may arise from a misinterpretation of the three likely scattering nodes of the electron.)

 Aerospace engineers probably have been aware since about the 1940s that the limit of "c" for any vectorial velocity-- incorporated through Minkowski Space into Einstein's Space-Time Model--was in error. It appears to be somewhat more than a rumor that "hyperlight drives" have been known for some time, but navigation and communication problems exist. *Since a message takes up a finite portion of a carrier wave, it can be shown that, for an intelligent message to be transmitted and received, the fastest possible relative velocity between the transmitter and receiver pair is 0.7...($0.5^{0.5}$) of the carrier-wave velocity. [This is for a transmitter receiver pair in orbit to one another. For other situations the problem appears at about half the carrier wave velocity, or even less.]* It may be, however, that there have been drives operating at about 0.2c for some time. These drives-- apparently utilizing superconducting at space temperatures--would be fast enough to make trips to check on a Mars Base reasonably feasible; but, would still be slow enough to utilize electronic technology for communication and control. .

The model suggests the possibility of molding super conducteurs into forms rather than needing to do windings. It is possible, for instance, that a left- handed Mobius Strip of superconductor--mounted correctly with respect to a permanent magnet in a "vacuum--" could be used to "extract energy from the void." A Mobius Rim (a double-sided Mobius Strip made by twisting twice in the same direction before pasting the ends together and then turning it inside out) might work even better. *Possibly the developing Graphene Technology could be pertinent.*

All Matter units may originate from the "Zerotron" unit which arises from the combination of the electron form and the anti-electron form into a single oscillator. This is a process which has been erroneously called "annihilation" . A shock wave could distort this oscillator to the neutron which subsequently collapses to the electron and proton. These recombine into Hydrogen atoms. Further elements may be built up from pressure effects on what we would call "Diatomic Molecular Cations."

Examples would be, HH+ compressed to D+, DD+ to He4+. and DH+ to T+. This last unit, the Tritium Cation, appears to be interconvertible to the isomeric Helium Three Cation, He3+.

It can be shown, mathematically, that any isotope above He4 may be considered as made up of combinations of D. T and He3 Units. Also, it appears that most, if not all, Beta Particle Emission, both negative and positive, and "K-capture" changes, are explainable by Tritium/ He3 Interconversion within atoms.

 This suggests that those radioactive decay processes mentioned above, may be through a mono-cation species and that Radiochemists may find that half-lives can be influenced by chemical environment. The instability of the Be8 unit may not be of the Be8 atom as a neutral unit but of the Be8++, dication, whcn is a perfect "set-up" to split into the units of He4 and "Anti-Alpha," or "Anti-Helium 4," and Alpha Particles. *[We probably should not consider the Alpha Particle as a He4 Nucleus but as a quite stable unit containing 6/8 of the He4 atom.]* Many of the ideas of Molecular Chemistry may apply to atoms. For instance, the idea of "Resonance Stabilization," from Organic Chemistry, fits well with the stability of the Alpha Particle as a "Square-planar Array." Alpha Emission may be a function of the dication form of the emitting unit. Analysis of atoms from the view of internal units, or possible units-- including T, He3. D and Be8-- may be useful in correlating the various breakdown patterns of radioactive atoms.

The Basic Force as an all-surrounding pressure, pushes interest at the atomic level toward the ideal of spherical symmetry. Anything that interferes with the nearly spherical symmetry of a diatomic unit, e.g. , HH, ,may cause it to tumble or spin erratically, This type of motion allows loss of "vib-rot energy" to the milieu and compression toward a spherical, atomic form. It may be that HH need only go to a triplet state to start the compression to D.

(The Deuterium Atom may be considered as a close

analogue of a triplet state.)

If this be true, it may be possible to cause the transformation of HH to D by the use of light of an appropriate frequency. The same would hold for DD to He4. However, it would seem that a DD+, cation would not be difficult to induce in the same way. This cation could possibly change to He4+ more easily than could the excited, but unionized, and hence more complex, triplet state to He4.

This idea of varying stabilities of ionic states and the possible change of diatomic ions to the monatomic forms suggests that there may be an interesting cycle going on in suns on their way from He4 to C12. One probable intermediate is Be8 which we have mentioned before as probably breaking down through the dication. it is possible to write an interesting sequence of reactions running a cycle around and around among the Alpha Particle, Helium Four and the Be8 ions which could continue indefinitely until broken up by an interfering reaction leading to C12. This "Chemical Perpetual Motion Machine" would be actually extracting energy from the surrounding void. If the postulation be valid, it would be interesting to try to "bring it down to Earth."

Very cautiously!

The area known as "Cold Fusion, " which concerns

the transformations at the subatomic level that can occur under less than the highly energetic conditions of plasma fusion, would seem to benefit from the ideas of this simple approach. The "One Force" is going to be most noticeable at the surface of a unit, where the imbalance of motion will be the greatest. In the situation of the formation of He4 on the surfaces of Palladium electrodes, it would appear that the saturation of the Palladium with Deuterium, probably as DD molecules, would furnish a constant feed of DD to a surface layer, where the DD molecules would be available to enter into a chain reaction situation if some unit could be produced to initiate the chain.

A DD+ unit could be produced by a number of possible initiating conditions. Two of these are the presence of an Alpha emitter and light of a sufficiently high frequency. Other possible initiators would be a polarity reversing pulse, and contaminants having initiating ability. A possibility that comes to mind is Magnetite, Ferrous Ferric Oxide--quite stable and magnetic.

By the above view, electrolysis furnishes the Deuterium but may not be necessary to the further processes. It would appear that it be possible to produce the Deuterium elsewhere and have the reaction take place on dry catalytic surfaces. In the ultimate, one might visualize a reactor consisting of a chamber wherein simple HH were transformed on one catalytic surface to DD which in another chamber would be transformed to He4. If the speculation of there being a possible He4 to Be8 and back "chemical perpetual motion machine" were to be shown to be true, then He4 would be used in a separate reactor to "extract heat from the void."

(Some people are bothered by the lack of observation in "Cold Fusion" work of the various electromagnetic emissions that are associated with "nuclear" reactions, i.e. the emissions that are usually associated with the formation or transformation of atoms. One factor overlooked is that the possible range of frequencies of our universe may run from a high frequency cut-off of "1/h" to a frequency of "ch" at the low end. Both of these are many orders of magnitude from the limited range which we can detect.)

The basic equation which arises from setting Planck's Constant equal to its definition as an angular momentum and assuming that the average tangential velocity is the Speed of Light, "c." is "m x r = h/c."

This little equation, which starts from the same basic information as do Quantum Mechanics and Space-Time, seems to be somewhat more inclusive than its better known "relatives."

In the expanded form of the m x r = h/c equation which is used in the basic information of the Oscillators-in-a-Substance Model, which is, as follows:

$$|Am| \times |Br| = |ABmr| = |h/c| = |Bm| \times |Ar|$$

--where |A| is the absolute numerical value of the rest mass of a particular oscillator in the cgs system, and |B| is the absolute numerical value of the corresponding Compton Wavelength in the same system--the equation could give rise to a new field of mathematical research dealing with the interaction and packing of the two different "size" vortex oscillators known as "electrons" and "protons." A "packing" which gives rise to the portion of reality which we call "Matter."

The use of absolute numbers avoids the problems inherent to the usual convention of a number being "positive" if there be no designated sign. That convention accidentally places most mathematical calculations in the right-upper-forward octant of a Cartesian-coordinate system. A mathematical model of an oscillator, however, is not compatible with the use of but one part of the available "mathematical space." Hence the use of absolute values, with the equals sign representing balance as well as mathematical equality.

The implied "Theorem" used in "O/S" to determine the second (high mass/small radius) limit for oscillators and the inversion values could be stated as follows: "If there can be found in nature a relationship of the form, xy=K, one may postulate that there is a reciprocal relationship wherein the absolute coefficients of x and y may be interchanged within the units system being utilized. Additionally there will be a set of values corresponding to the square root of the absolute value of K which will correspond to the inversion values of an oscillator." Several interesting ideas arise.

One of these ideas starts from the fact that "c," the Speed of Light, is defined as "wavelength times frequency, c=hNu". By the above theorem, "c" could be considered the constant, K, of a set of oscillators which would show, with separation from the transmitter, a continuing redshift for initial broadcast frequencies above the value corresponding to the frequency-wavelength defined by "$c^{0.5}$." This would be expected.

However, this theorem suggests the surprising idea

that *frequencies below the above square root values would show a continuing blue shift with distance, which seems totally contrary to logic.* In, addition, it suggests that there would be a frequency/wavelength corresponding exactly to "$c^{0.5}$" which would be *stable over distance!* This frequency happens to be within a range which can be studied. (It is about 173 kilocycles/sec.) There could be some sort of a possibly utilizable "motion equilibrium" at this value.

It was not until 2014 that it was noticed that the expression "h/c" would actually be associated with a time-independent value, per cycle, and would; therefore, be a basic quantum of work, thus giving additional validity to the idea of a "Basic Control Oscillator." The following little article resulted. *This article was written some six months after the Infinite Energy article. The very deep significance of the h/c ratio was not realized until there was a discussion of Charles William Johnson's analysis of the Speed of Light as an average, published on the Earth/matriX Web site where it was realized that "a wave length was a wavelength was a wavelength, " That is, the Energy/Work content of a wavelength is independent of its duration.*

THE BASIC QUANTUM OF EXISTENCE--IN PLAIN SIGHT FOR A CENTURY

Summary: An overlooked ratio appears to be a measure of the smallest "Quantum" of Existence.

A look at the dimensions of the Constant of Nature, " h/c" snows it to be a time-independent constant having the value of mass times distance per cycle., independent of the length of the cycle as compared to any other reference cycle. (this last statement is simply emphasizing that the value is independent of Time),this may be considered as the basic "Quantum of Existence." That is, the basic unit of work which could be said to have given its name to the "Quantum Revolution" of the 20th Century.

While "Purists" will disagree with the statement, "Mass times distance equals work," claiming that the definition of "Work" is "Force times distance," this writer contends that any measurable mass will alway have associated with it a basic unit of "acceleration." so that, for general discussion, it is immaterial whether we say "Mass" or "Force," contending that the measurement of "Mass" is a measurement of the "Basic Force" involved in whatever instance is being considered.

The little quantity, I h/c I, which-- in words--is "The

absolute value of Planck's- Constant-divided -by -the - Speed-of-Light," could be the "Work" involved in rotating some basic unit of existence once around.

This value-- about 2.2 x 10^{-37} gram-centimeter-- in the cgs system--also, would be the amount of work involved in any wavelength of electromagnetic energy, no matter what the wavelength be. That is, the greater the wavelength, the lesser the mass value per unit of distance and vice versa.

Planck's Constant would be the value associated with rotating some basic--or, perhaps, average--unit at a tangential velocity of the "Speed of LIght."

In published work on the Oscillators-in-a-Substance Model. the use of this constant, h/c, as the balance constant for an "Oscillator Family" has the effect of showing that this little work function is applicable to an unlimited set of possible motion combinations. Among those combinations are the electron/positron unit and the proton/antiproton unit as well as the neutron and--probably the most important of all--, a possible "Control Oscillator of Our Universe" operating at the mass and radius values corresponding to $(h/c)^{0.5}$-- about 4.7 x 10^{-19} cm. wavelength.

A high frequency oscillator of this type operating continuously into a 'paleo-substance," converting that into a "proto-substance," (e.g. a unit such as the postulated "Zerotron,") which would be splittable to the electron and positron and distortable to the neutron, This proto-substance would be forced to expand outward as more and more accumulated. Considered in the light of new information about the microwave background and where it is clustered, it is not too hard to explain "Black Energy," "Black Matter," and a Basic Origin Instant in terms of this one high frequency, continuously operating oscillator. In other words, the "Big Bang Instant" would not have been some sort of one shot from a "Cosmic Cannon," but more like the start up of a continuously operating "Cosmic Machine Gun...."

PART TWO. A COLLECTION OF ESSAYS WHICH ARE NOT ESSENTIAL, BUT MAY BE INTERESTING.

IT IS CONSIDERED THAT THE BASIC MATERIAL OF THE MODEL HAS BEEN COVERED IN THE PREVIOUS PAPERS. THE REMAINDER OF THIS BOOK RANGES SOMEWHAT FURTHER AFIELD INCLUDING REPRINTED ARTICLES THAT ARE ESSENTIALLY EDITORIALIZING. WITH COMMENTS AS TO WEAKNESSES AND PROBLEMS OF THE OLDER MODELS, ESPECIALLY WITH MATHEMATICAL ASSUMPTIONS THAT MAY NOT COINCIDE WITH

PHYSICAL REALITY.
THERE IS AN INTRODUCTION TO USE OF
IMPLICATIONS OF THE MODEL IN THE FIELD OF
NUCLEAR CHEMISTRY, CONSIDERING THAT FIELD
AS AN EXTENSION OF MOLECULAR CHEMISTRY.

The following two papers, published as one unit on the Earth/matriX Forum, probably have had the widest dissemination of any of the papers. Written in 2008, they are generally valid, however, there are nuances in later papers. The Internet references included are possibly no longer valid due to Google changes.

-

Oscillator/Substance Model View of Elements and the Periodic Chart
By Dean L. Sinclair

Atoms are conventionally considered as being made up of a nucleus consisting of protons and neutrons surrounded by an "electron cloud." These atoms can be arranged in groups known as elements and these elements can be arranged into a "Periodic Table." This way of considering things in considered basic to chemistry.

A new model of reality suggests, however, that useful as those concepts have been, they may be somewhat in error. This model, called the "Oscillator/Substance Model considers all of existence to be within a "substance" made up of oscillators and/or controlled by oscillators

If the Oscillator/Substance concept of what are electrons and protons is valid, and electrons and protons are actually spinning vortices, with the size of the protons and their interactions accounting for the "nuclei" of the nuclear atom, then it is necessary to do a complete re-examination of the concepts of atoms, elements and the basis of the Periodic Chart of the Elements. A new reason/rationale need be developed for the apparent presence of "isotopes" which are conventionally explained by the presence of neutrons in atomic nuclei. If there be no neutrons per se in the nuclei, then, what are the electron-proton interactions which explain a situation which will make it appear that there are stable neutron type associations?

Conventional wisdom says that there is a pairing of electrons such that their "spins" will cancel. O/S would agree that a partial cancellation of the vortex motions would take place if two vortices were "paired, upside down to one another." The conventional literature, however, does not continue to apply the same idea to protons, nor does it seem to realize that an up-down-up-down chain or circle can be extended. It should be useful to consider the first few known proton-electron associations.

The simplest units which may be considered as proton-electron associations are the neutron and the Hydrogen atom, H1. The neutron may possibly be [or--at least, can be thought of] a tightly coupled dual oscillator of an electron and proton, or proto-electron/proto-proton. And, when it "falls out of sync" it collapses into an electron and a proton. These two can reunite into another unit which can synchronize in unlimited --or, nearly unlimited--- spatial dimension. Hence, it cannot be "knocked out of sync" in the manner of its more restricted isomer, the neutron. This unit is, of course, the H1 atom.

Perhaps the next easiest set of units to consider is what one might call "Iso-set--2,2," the set of units made up of two electrons and two protons. This would have also two members, the Deuterium Atom, "D," or, "H2," and the Hydrogen molecule, H:H. The Hydrogen molecule is known to have two forms known as "*ortho*-Hydrogen" and "*para*-Hydrogen" wherein the Hydrogen nuclei, i.e., the protons are "matched spin" or "paired-spins." In other words, the protons are moving in exactly the same orientation in one case and are "upside down to each other" in the more stable orientation. The Hydrogen molecule would be expected to be an ovoid. It is probable that the Deuterium atom could be considered as a condensed version of the "*para*-Hydrogen" molecule, containing much less vibration energy.

The close-coupling of the two protons in the nucleus and their probable containment, at any given instant within an electron, give the illusion of a "nucleus" consisting of a proton and a neutron. It can be considered that the vibration motions of the electron and the protons would be such that for very short periods, the proton/electron interactions would be such as to be identical to those of a neutron, however, the neutron as such would not exist.

The three electron / three proton set, "Iso-set--3,3" has two well-known, atomic members, Tritium, "H3," and Helium 3, "He 3." Tritium is "radioactive," spinning off an electron to form a cation which as it regains an electron becomes the isomeric He3. Tritium does not have a magnetic moment; therefore, it is not a "spinning neutral unit." This implies an internal symmetry when the nucleus is composed of an inverting tetrahedral array. This is represented by three units inverting as if they were at corners of a tetrahedron; this unit then would on the average have no polarization, no "dipole", even though the internal units have inherent dipoles. It appears from the chemistry of this unit that, at any given instant, the central array or the "atomic nucleus" is encased within a set of two coupled electrons with a third electron more loosely coupled.

The "stable" He3 unit is a totally different configuration. With a definite magnetic moment in the same sense as that of the neutron only larger, it is a spinning "neutral" atom with a definite dipole dominated by the spinning protons. It would thus appear to be in a trigonal array corresponding on the atomic scale to the "resonance-stabilized" trigonal units that are known in molecular chemistry. These three appear at any given instant to be tightly encased in the vibration pattern of one electron and more loosely in the vibrations of two others. This gives rise to the conventional idea of one neutron in the nucleus. Evanescent, transient states which would be identical to states of neutrons, would exist in both the Tritium and Helium 3 units, as well as in all atoms, giving rise to the illusion of stable neutrons in the nuclei.

In the "Iso-4,4-set," it interesting to compare the molecular unit, Deuterium molecule, D:D, and its atomic isomer, the He4 atom. This has been covered in some detail in another short paper, "*O/S Theory and the Deuterium to Helium 4 Transform.*" In that paper it is noted that the Deuterium Atom nucleus consists of two, "coupled-up-down" protons (in the most stable state) and the Deuterium Molecule would be two sets in what could be called a "stretched tetrahedral array" or possibly a "stretched square planar array" or, an arrangement combining these two ideas. This set of "nuclei" would be surrounded literally within and without by an array of 4 electrons. The overall result is an ovoid with two distinct centers of motion.

The He4 nucleus would be composed of the same eight basic units. However, in this case, the ovoid of two distinct centers of motion is compressed into a spheroid having but one center. It is possible that this spheroid would have a tetrahedral configuration of protons, again surrounded, within and without, by a tetrahedral arrangement of electrons.

The Alpha particle would not be a He4 nucleus, but a square-planar (actually, circular) array of 4 coupled protons surrounded, again, "within and without," by two electrons. The coupling of the protons gives the unit a very strong, possibly clockwise spin.

This imagery makes sense if one looks at the oscillator limits of the proton and the electron and the reversed spin/inversion senses which give rise to the "positive" and "negative" charges.

The electron and the proton by Oscillator/Substance logic have the same "average" mass and radius. Nonetheless, they have very different oscillation limits, with the electron limits approximately "ten to the third" times those of the proton. Hence, both have the same rotational/inversion velocity of "c"; the proton completes many inversion/rotations for each one of those pertaining to the electron. The electron is therefore "heavier and lighter" and "larger and smaller" than the proton.

One may observe that the analysis of the entire periodic chart and corresponding molecular isomers would constitute a difficult if not impossible task. It appears that Nature is such that certain patterns arise as of electron-proton interactions which lead to the Periodic Chart pattern with the use of "*neutrons in the nucleus*" as a convenient bookkeeping tool. This must be kept in mind if one feels that the O/S insights are valid. The "neutrons in the nucleus' should be considered illusional and it can obscure other patterns of value. Some of these can be found in the use of the Iso-set | Iso-A concepts covered to some extent in a previous paper: "*Iso-sets, a Key to Radioactivity*." That paper may be found on SciScoop, and as a web-page on the *Google Group, Oscillator/Substance Theory*.

Note added March. 2015. The following paper is also from the Earth/matriX Forum where it is published as an addendum to the above paper. The original publication was in Sci-Scoop. as of the 11/17/08 date noted. Hence this is an early paper that may be of historical value to the model .

A Constant's Secrets. A Different Look at Planck's Constant

Physics Monday, November 17, 2008

For the last hundred years, in plain sight, there has been hidden within a Constant of Nature, important information about the Universe within which we live. Planck's Constant was discovered in the early 1900 by Max Planck. In all the time since, however, it appears that no one has taken the time to ask some logical questions about the Constant. Some of these questions are the following: The constant applies to the relationship between Energy and frequency of electromagnetic radiation, so what is it operating on to connect Energy and frequency of radiation? Why does it have the dimensions of "Action" or "Angular Momentum?" It is logical to ask, " On what is the action occurring? What has the angular momentum?" Running these questions around brings to mind the thought that if this factor, Planck's Constant were an average amount of action in any direction, or the average angular momentum in any direction— apparently the same— of units making up reality, then that average unit would appear as a constant.

Planck's Constant is used at the speed of light, another Constant of Nature. Could the thought that perhaps Planck's Constant is an average value of a characteristic of some unit of existence also apply to the speed of light? Could the Speed of Light also be connected to something related to rotating units as Planck's Constant may well be?

This last speculation fits in very well. The Speed of Light is the maximum velocity of information transfer. Information transfer in any direction from a point has a maximum–and in practice, unattainable–velocity of the average velocity in any given direction of the velocity of the "packet carriers" whether these packet carriers be Pony Express Riders, Sound Waves or Electromagnetic Radiation. The Speed of Light would make sense as the average velocity in any direction at any time of the units, rotors(?), acted upon by Planck's Constant.

Acting on these ideas, Planck's Constant is an average angular momentum and the Speed of Light an average velocity in any direction, we can try some mathematical analysis.

Let us set Planck's Constant, (h), as the constant value into the form equation, K=xyz. In this case, let Planck's Constant b "K", "x" will be an instantaneous mass, "m," rotating at a distance, "r." from the center of rotation, and the third unknown be an instantaneous velocity, "v" measured at that distance, "The Tangential Velocity." We write, "h=mvr." Since "he" appears to be necessarily valid only at the speed of light, "c," we evaluate h=mvc. Since "c" itself is a constant, the ratio of "h" and "c" is a constant so mr=h/c. This two unknown equation can be said to define not only a variable area but also anything which varies in this manner. One such is the Law of Levers in Physics. The numerical values of "m" and "v" can be interchanged, hence this little equation could describe something moving back and forth between two limits or two "states." Such motion is called "oscillation" and something acting in such a way is called an oscillator. We can say that the equation, mr=h/c, describes a family of oscillators. Since the values of m and r can be switched, they can also be equal to define an "average" oscillator when $m=r=(h/v)^{0.5}$. That is, when the numerical value of the mass equals the numerical value of the radius and each equals the square root of the ratio of Planck's Constant to the Speed of Light. Inserting the value of 6.63×10^{-27} erg. sec. for Planck's Constant and 3×10^{10} cm./sec for the Speed of Light we get a value of approximately 4.7×10^{-19} cm. (and 4.7×10^{-19} grams) for the "dimensions" of the average oscillator defined by the "$(h/c)^{0.5}$ Constant."

It is interesting that this value, 4.7×10^{-19} cm. as a radius, is almost exactly one half of the diameter at which the Strings of String Theory are said to disappear into a "10 dimensional hole." Also, Quantum Mechanics is said to fail at below the same distance.

At this point we suggest that the basic units of our existence are tiny oscillators half of whose existenceS is unknown and unexplored as it lies below the "threshold" of $(h/c)^{0.5}$.

Perhaps we should close this little essay at this point and hope that the reader's curiosity has been aroused enough to follow up. There is much more that can be said and there has been much more written elsewhere.

An Internet Group, Google Group, Oscillator/Substance Theory has been set up to explore, develop, publicize, or refute, the "Theory of Everything" which arises from the foregoing and related topics. It is open to anyone, and anyone would undoubtedly be able to contribute. The URL is as follows:

http://groups.google.com/group/oscillatorsubstance-theory Hope to see you there!

Dean L. Sinclair, Aberdeen, SD Nov. 2008

The following essay is focused on the fact that the

basic ideas were available a long time ago and could have been of value if seen much earlier. It is more of an editorial than it is scientific writing.

A Century of Bumbling

The simple Oscillators-in-a-Substance Model of Existence, which seems to give logical answers to physical science questions except for the fact of Existence itself, could have been developed a century ago instead of within the last decade. The basic data was there. The Michelson Morley Experiment had defined the speed of light as a constant. Planck had defined his constant.

Indeed, basic theory based on the concepts of Albert Einstein did develop from this data. Unfortunately, these concepts were flawed and led to serious misconceptions.

To make things worse, Isaac Newton had considered Gravity to be an attractive force between matter units, apparently not realizing that this would not be a "True Force" - one having an "Equal and Opposite."

Had he realized and noted that Gravity could be formulated as a push-together, a pressure from some all-surrounding substance, then the key insight would have been available in his day.

One may, if they wish, blame this error by one of history's geniuses for subsequent errors in logic of Einstein and subsequent physicists, who, to this day, through a "Century of Bumbling," filled with misconceptions, have left the physical sciences in a tangled mess of "accepted science," much of which, on analysis from the "O/S" view, is arrant nonsense.

The multi-billion dollar Hadron Collider built at CERN under Switzerland and France, was conceived to find the "Higgs Boson.".The Higgs Boson is, supposedly, a transitory- - or possibly ever-present -- unit which gives "Gravity" to units made up of "massless" quarks, gluons and gravitons.
To be blunt, the collider was looking for a phantom based on nonsense. In 2013, it was announced that a "Higgs Boson" had been found.

Cynically, one may suggest that something that could be called a "Higgs Boson" had to be found to justify the expense. Probably, the scientists didn't "cook the data.", They found something, and interpreted what they found as what they wanted to find.
"O/S" suggests that if one started with the "electron/positron" entity and continuously

accelerated it in an "electromagnetic field", it would alternatively increase in velocity as the field pushed on it vectorially, and decrease in size (increase in mass) toward the "average", inversion size of about 4.7×10^{-19} cm. in radius, apparently passing through a proton and antiproton stage and several other stages of closeness almost to "ultimate compression" at the 4.7×10^{-19} stage.

The Hadron Collider people -- the scientists at CERN -- may well have found another stage, previously unknown; but, they didn't find a "Higgs Boson. ".

In a sense, the "entity" at radius of about 4.7×10^{-19} cm. is either always there, or only instantaneously there, in every oscillator. might be said to be sort of "Higgs Boson" of "O/S.".

In a way, there is another "Higgs Boson of O/S," a central basic unit. The "Zerotron," which is the unit in "O/S" which "electron-positron" units combine into and which can be split back to them or deformed to neutrons.

This unit appears to be the "Parent of all Matter," and may be the main occupant of all "voids." This unit would be the unseen occupier of the ignored "Null Set" in any process.

If it be the basic unit of the "Substance," the Zerotron being distorted to neutrons may account for the microwave background of the Cosmos.

As the shock wave generated through the "Substance" by the activity of the oscillator that generates Our Universe distorts Zerotrons, the initial forms shake down to the "stable" neutron, producing microwave frequency disturbances.

These microwaves would show a cone-shaped pattern reflecting the course of the shock wave, with a continuation toward the point of inversion of the oscillator; that is the point in "Space time" of the "Inversion Instant," when the Control Oscillator came into existence and Our Particular Universe began.

Recent mapping of the microwave background seems to agree with this assessment.

This could probably go on to a book-length essay covering more of the "mysteries" engendered by conceptual error.

"Black Holes," "Black Matter and Dark Energy", "Matter/Anti-matter Annihilation", "No relative velocity greater than light is possible" are a few of the accepted misconceptions.

"Space-time" turns out to be a fairly valid

mathematical model of a substance, but without oscillators. At the other end, QM and String Theory have oscillator resemblances.

It may be noted that String Theory does fill the "void" with something.

There is a definite possibility that-- had Science followed a reverse pattern, as suggested here-- we would have had intra-solar-system travel developed and usable before the time of the wasteful "Space Race."
That travel would have been using concepts which were apparently already known in the 1940s. That is, concepts of drives capable of near-- or even hyper--light velocities. Drives which were "impossible" -- and still are -- by conventional accepted "Science." Drives,that, however, are entirely logical in the O/S vein.

Yes, we made great progress in the 1900's. We almost certainly could have done much more had we not bungled physical science theory badly, forcing technologists to go ahead and do "what worked," with no idea of " why or how."

The following, posted on Helium,com, is one of several articles of the same name--according to the policy of that site. It could be said to be "Blatant heresy, with respect to the scientific establishment. " It makes some of the same points as the article above as to the changes that this model make in the consideration of reality. It predates the above article.

The Large Hadron Collider: Greatest Physics Experiment in the World

Introductory summary: This writer takes the view that, although the Hadron Collider is undoubtedly the "Greatest Physics Experiment in the World," it may also go down in History as the most expensive colossal failure of an experiment ever attempted. The writer is of the opinion that because the scientific premises of the experiments that it is designed to carry out are flawed, the Collider will not operate as expected.

The lead article of this title, does a magnificent job of describing the Collider and the expectations of what it was designed to do. However, the scientific premises for the experiments to be carried out are based on very shaky grounds. The theoretical basis of the experiments to be carried out are based on the "Standard Model of Particle Physics," which received a Nobel Prize in Physics in the 1970's and has been cited as "The Crowning Achievement of Particle Physics." Despite these impressive credentials, the Model is based on very shaky premises and is coming under strong questioning.

The Standard Model considers that there are Four Forces of Nature, Gravitation, Electromagnetism, the Strong Nuclear Force, and the Weak Nuclear Force. It then decides that Gravitation has negligible effect at the atomic level and can be ignored. Additionally, it is assumed that the units created in atom-smashing experiments are fundamental units, somehow released. A further assumption is that units called Quarks are even more fundamental particles.

A long paper could be written on the above points. This will try to be a brief summary of some of the problems with these ideas. None of the "Four Forces of Nature" meets the definition of a Force. That is, None has an equal and opposite reactive Force. Electromagnetism and Gravitation describe sets of observations which can be described mathematically; but, they are not "Forces." The two "Nuclear Forces" appear to be totally imaginary constructions which are necessitated by the Proton-Neutron Model of Atomic Nuclei. This model, itself, although it has been accepted since the 1930's, has some logical inconsistencies and is, itself, being questioned.

As to the question of "Quarks--" When the writer, as a young graduate student some years ago, read the first paper that proposed the Quark, the strong impression came through that --although the paper was presented "Straight faced--" it was actually a satirical "send-up" on the tendency of Science to complicate things with increasingly more esoteric and incomprehensible models. Although the idea has been taken seriously for many years, the first impression still seems, to this writer, to be a real possibility. The electron scattering data which has been considered as proving the existence of "Quarks," may be simply a misinterpretation caused by a lack of knowledge of the oscillatory nature of the electron and proton.

The construction of the Hadron Collider, designed primarily to smash protons into protons to prove ideas from a theory whose base is very shaky--to say the least--probably dooms its primary experimental premise to continual failure.
This is particularly true if insights from a relatively new model, the Oscillator/Substance Model be correct.

-

The "O/S" Model is a simple-in-principle, but very

comprehensive, model which arises from reinterpretation of basic data, some of which is over a century old. It considers existence as being within a Substance/Substrate which is controlled-by/consisting-of oscillators. From this model arises descriptions of the electron and proton as "rotating. inverting vortexes" of grossly different sizes--the electron inverting through a radius 1832 times that of the proton--and having opposite rotation/inversion senses. This model also contends that any "vacuum" actually is occupied by the "Substance/Substrate." and, furthermore, that a rather common unit within this substrate will likely be a spherical, pulsating oscillator, dubbed the "Zerotron"--as the zero point from which the negatron and positron diverge--which is splittable to the aforementioned electron-anti-electron pair and deformable into a neutron.

If this model be valid, it can be seen that pushing a spinning vortex--which can couple with like vortexes--through some sort of a medium is a far different situation from lining up positive particles--which will mutually repulse--and firing them through a void. The O/S Model suggests that, as soon as they try to get their machine up to any real degree of power, the operators have a "tiger by the tail," which they do not realize exists. The coupling of protons to proton, the possible formation of electron and anti-electrons from the "medium" suggests that--instead of having a simple accelerator--very soon after startup, the machine will operate as a gas-fusion reactor. "Energy feedback" is inevitable. Considering the above, it is no wonder that the Hadron Machine appears to have a history of almost immediate breakdown.

-

There is good reason to believe that a stream of protons entering into an atmosphere of Hydrogen Molecules, could create Hydrogen Molecular Mono-cations which would "spin-down" to Deuterons. forming Deuterium, which could couple with protons to form Tritons and, thence, Helium 3, and couple with deuterons to "spin down" to a Helium Four Cation. These processes would probably be close to what happens in the early formation of stars from clouds of Hydrogen gas. Not only would studies of these gas phase reactions have scientific value, but,also, would be of great economic interest from the "Energy" release.

-

A possibly optimistic thought: "Perhaps the Hadron could have a useful future, if it were modified to be an experimental gas-fusion reactor, exploring the early processes of star formation, rather than trying to continue with the project of finding the 'Higgs Boson,' which is a 'will-o-the-wisp' based upon questionable "Science."

Additional note, Jan. 2015. The scientists did find "something" which they said to be the "Higgs Boson." Higgs and his "co-authors" got a Nobel Prize. The writer is still unconvinced, and it appears that he isn't the only one.

Incidentally, it appears that the Collider may have been modified somewhat after the first breakdown, Also, it appears that there was care to keep the protons separated except when on collision course. As the Oscillators-in-a-Substance Model gives an easy, logical explanation for Gravity, the idea of a "Particle which bestows Gravity upon other particles" appears to be total nonsense,

There was another early paper on Helium.com under the title, "Why Einctein was Right," which was not exactly respectful of an "almost-sainted" scientist. Although not focused there, it does imply that the "Mathematics-is-the-reality" attitude taken generally by Einstein and the theorists of the intervening years may be misleading.

-

Why Einstein was right, when he was, in his

"Relativity" theorizing, was often because hidden facets of mathematics and of the definitions used in physics allowed the ideas to be usable, even if his reasonings were fallacious. In other words, often, he was lucky.

-

Einstein's Special Relativity, which title for it I am told he disliked, describes how information is changed as the relative velocity of a transmitter/receiver pair approaches the velocity of the carrier wave. When the carrier wave velocity is taken to be the limiting relative velocity of any two independently moving objects, the Special Relativity view leads to nonsensical conclusions.

-

"SR" accurately predicts that "Mass" will increase when a moving object reaches the Speed of Light. However, the prediction is that the Mass will go to "Infinity." There are some problems with that. Infinity, in practice, simply means that our measurement device, or our logic, fails at this point. There are two other factors, of which Einstein seems to have been unaware, which come into play here to allow his model to accurately predict a change in the situation at the Speed of Light.

-

First, mathematics does not allow empty space. There is always an implied dot field or dot matrix, so it can be expected that whatever is considered to be moving would be moving within "something."

-

The other factor is also mathematical. When the process of "Integration " is carried out on the momentum equation, (m x v = p), mass times velocity equals momentum, this process may be carried out with either the mass or the velocity considered as the variable. If velocity varies, we get the usual Kinetic Energy Equation, $KE=(mv^2)/2$. However, if mass be considered as the variable, with velocity constant, we obtain another energy equation, $(vm^2)/2$. This latter does not appear in the literature and has apparently been ignored. These two equations may be interpreted as indicating that "velocity" can change to a limit. If the velocity hits a limit, and the "accelerating situation" continues, "mass" will change. Neither the "dot-matrix" aspect of mathematics, nor the alternative energy formulation, seems to appear in any of Einstein"s work, nor anywhere else in the readily available literature. If both the dot-matrix and the alternative energy expression that appear in the mathematics are reflected in reality, we see that any moving entity within the matrix will, itself, be a part of the matrix and the mass will be a measure of the balance of the interior of the moving entity and the remainder of the matrix. As long as the translational velocity of the moving object is small with respect to the average speed of motion in the matrix, there will be little effect of the "second" energy equation. Velocity will change and the amount of disturbance dissipated by the matrix, i.e., the "Energy" will be measured quite accurately by the "Kinetic Energy" formula. There is a change in the situation when the translational velocity of the moving object starts to equal or to exceed the average of the matrix.

(In the "Universe" In which we exist, this average is known as "c," a Constant of Nature which is the Speed of Light in a vacuum.)

At this point there will be a significant change in the balance between the rest of the matrix and the part of the matrix within the surface of the moving entity. This surface will become changed in size and shape. The motion disturbance is no longer primarily dissipated into the matrix at large, but becomes localized at the surface of the involved entity. The balance changes, the "Mass" increases. Special Relativity turns out to have been at least partially correct. The situation changes at the Speed of Light.

A third factor hidden in the interface between mathematics and reality apparently allows Einsteinian

Space-Time modelling to give quite accurate predictions. This is in the definition of "Time." In the "Space-Time World" of conventional thinking, "Time" is a reality which somehow came into being at the beginning of "Existence." Actually, Time is a convenient method of keeping track of motion in sequence by a measured interval. The hidden factor that apparently makes "Space-Time" modelling work is that "Time" is always referenced to some reproducible cycle in nature. Therefore, "Time" has hidden within it not only the idea of motion, but also the idea of cyclic motion. A unit of time, a second, for instance, can therefore represent a cyclic motion, or motion in a circle, and the expression "sec.^{2}," could stand for the motion content in the volume of a sphere! This insight leads to some interesting interpretations of some of the equations of physics, which, unfortunately, are beyond the scope of this paper.

-

The "hidden factor in Time." however, makes Space-Time Modelling and Motion in a Matrix Modelling reach much the same conclusions and, at this point, they seem to be essentially equivalent approaches. The Motion in a Matrix users being, perhaps, more cognizant of why their ideas have validity. This "circle/sphere" aspect of time hints at the idea of a spherical oscillator as a basic entity. This latter idea, has become the basis of a variant of Motion in a Matrix Modelling which could be called the "Oscillator Substance Model."

-

The "Oscillator Substance" model, although developed independently, echoes Max Planck's ideas of dots controlled by oscillators and is the discovery of a chemist who was once trained as an electronic technician. who has, within the last year, put together insights from both fields to suggest that there is a very simple model of everything which perhaps would have been close to the "Unified Field Theory" which Einstein spent his life trying to develop but could not. We will come back to this later. First, however, since, we are focused on where Einstein was right, or wrong, we should note the errors that probably doomed Einstein's quest for a "Unified Field" from the start.

-

It is highly probable that the "First Fatal Error" in Einstein's Unified Field attempt was that he "Threw out the field." That is, he assumed a void, a nothingness, for the "Field" to operate in. The "Second Fatal Error" is one that modern theorists continue to make. They try to set up a unified field theory from the "Four Fundamental Forces." The problem is that the "Four Fundamental Forces" all violate the Law of Forces, "For each and every force there is an equal and opposite force." This law of forces, if examined carefully, can be interpreted to clearly indicate that any "Force" is simply a readjustment of pressures within a "substance." The "Four Fundamental Forces" are either, in two cases (Gravitation and Electromagnetism) descriptions of observed phenomena, which are the result of other factors, These kind of "Forces" are known as "Fictional Forces," the best known, and best explained of which is "Centrifugal Force." The other two "Fundamental Forces," the "Strong and Weak Nuclear Forces" appear to be simply imaginative explanations which arise as justification for the idea that neutrons exist, as such, in atomic nuclei. A much simpler explanation appears if one considers atomic nuclei to be electron-proton aggregates in which a neutron has, at the best, potential existence. Einstein's frustration was apparently caused by his operating on sets of erroneous ideas.

 In the next couple of paragraphs will appear a statement, which Einstein could have published, a logical outgrowth of Planck's ideas, Which might have kept us from almost a century of what this writer considers "semi-mystical nonsense" in scientific theory.

"Let us assume as a working hypothesis that there exists a 'dot matrix' of separable oscillators. These. in turn are collected to form a 'substance' at its 'triple-point,' of larger separable oscillator entities capable of correlating motion and of separation into the electron-anti-electron set and distortion into neutrons." Amplifications of the ideas in this statement, and corollaries, can give explanations for electrons, protons, the expanding universe, "The Big Bang" and almost any other phenomenon to which it has been applied. It is this working hypothesis which has led to the explanations of "Mass" and "Energy" which have been used throughout this paper.

Why was Einstein right? When he was, it was almost as much luck as brilliance. The hidden implications of the mathematics which matched reality made the theorizing seem to fit even when the basic ideas erred. Had he gone in a different direction, following up the ideas of Max Planck, he might have reached the same basic conclusions as the "Oscillator Substance Model" which has arrived nearly a century late, and will probably be ignored.

Postscript: Although this writer finds the Oscillator Substance approach so natural and useful as to feel that it should be common knowledge, the discovery of this possible "Explanation of Everything" is only a few months old. The ideas may wait some years longer for confirmation or disproof as what information published about it, thus far, is almost exclusively on Helium.com. and the discoverer has no professional standing as a member of any research institution or group. It is sad that a young, patent clerk could not have reached these ideas in the early 1900's rather than a very elderly janitor being the discoverer a century later....

Now, if a certain "String Theorist" were to have happened to have come up with this... Oh, well, we can't have a perfect world!

Einstein was very right to try to explain the workings of reality, it's sad that he made a few key mistakes.

There are several little papers which deal with simple mathematics as addressed in the Oscillators-in-a-Substance Model.

Reality and Mathematical Definitions

We consider mathematics as a model for reality. Perhaps it would be worthwhile to consider how certain mathematical definitions fit as we look at the physical world.

First, let us look at the concept of Zero, usually considered as the symbol of nothingness. However, is this true? Zero is the starting point for counting, the starting point for any journey, the crossing for the Cartesian axes of most conventional graphs. It might be better to say that the symbol, "Zero," is actually the symbol for the fact of existence. In the real world, the starting "point" may be of any size, any shape. Zero, then. is not without existence, without dimension, rather it is the symbol of the very first dimension, the Dimension of Existence.

What then of the number, "One? " That's simple, its the 'counting number." However, isn't it a lot more than that? It is the symbol of wholeness. it represents a whole starting point, a whole line, which actually has to be made up of two starting points, a whole surface, made up of at least three starting points, etc. Hence the number one may represent many things. If we attach a sign to the number, implying a motion, then +l, represents the motion of a whole starting unit one unit to the right, or possibly up, or forward. We say that one times one times one equals one, but we always assume that there is a positive value attached to the one, so if the first one represents a motion of one space to the right, the next one represents the motion of the first "one" upward, and the third one represents the motion of the second generated one a unit forward.   Therefore, if we attach the positive notation to "one" which, by convention we do, one times one times one actually means one cube generated to the right, above and forward from the origin of a set of axes by a set of motions which are actually counter clockwise,

Looking at "One" as the symbol of wholeness has many uses. One interesting one arises is one looks, for instance, at figuring a maximum frequency for our particular Universe. The equation for the movement of Energy by electromagnetic radiation is E=hu, where "h" is Planck's constant and "u" is cycles per some unit of time.   If we place E equal to one Energy unit in any set of units, the maximum frequency, expressed in that set of units, will be seen to be "1/h" . This presumably would be the "high frequency cut off" for communication."

At the other end of the scale is the symbol of "Infinity. the Number Beyond All Numbers." For mathematicians this is a perfectly good definition; but in the real world we have to use more rational definitions. Does it really make sense to say that we can measure mass of something moving with relation to us up to a velocity as near the Speed of Light as we care to but say that the mass will become "Infinite," meaning "without limit" at the Speed of Light? Is the darkness just beyond the flashlight beam a void? Couldn't that darkness be considered an Infinity? We can't see into it. In the practical world we probably should consider the concept of "Infinity" as representing simply the point just beyond the last point that we can measure with the instruments at hand, the number beyond where we stopped counting, for whatever reason.

This writer feels that, were the above, and similar ideas taken into consideration, that the theories of physical science would fit more closely to reality than do the current theories which are considered as best expressed in differential equations, which are always expressed in simple numbers, considered, of course, to have positive signs....

-

On Signed Numbers

The signs, + , and, - , are used throughout Mathematics and Physics in a number of ways, and with several meanings which are often not carefully checked. The plus or positive sign has its initial use in Addition in the sense of increasing a pile, of no particular dimensions, by a certain amount described by a counting number written after it. The negative sign represents the opposite operation of removing a certain specified amount.

In Physics, the positive sign represents a "charge" associated with the proton, the negative sign represents and "opposite" charge associated with the electron and other species having a characteristic in common with the electron. *[This writer suspects that characteristic is counter-clockwise spin.]* In this usage, the signs do not represent reversed operations but characteristics which are considered opposites.

-

A third usage shows up in Mathematics where the signs are associated with counting numbers to form sets of "signed numbers" which seem to be able to be added, subtracted, multiplied and divided like counting numbers. However, this turns out to have problems when one does multiplication and division processes. What is overlooked is that the addition of the sign to a number gives it both a magnitude and a direction. Something having both magnitude and direction is not a true number, it is what is called a "vector."

-

The signed number may represent movement away from a zero point, a line, a plane, a three dimensional figure formed of planes or some "higher order figure" depending on where it occurs in a sequence of operations. *Signed numbers are. handled according to a convention wherein the positive sign is considered as being to the right of an origin, upward from an origin or forward from an origin, and if one multiplies three positive signed numbers together, say plus two times plus two times plus 2, (+2 x + 2 x +2) what one has really described is moving two units to the right of the zero point, moving this "two-units-line" upward to form a square, then moving this square two units forward to create a cube which is situated to the right, above and in front of the origin point. This "eight-cubic-units entity" is called a positive volume because we say" + x + x + = +"* as we consider that the + sign represents travel " in the same direction" while the negative sign represents travel in the opposite direction. This signed unit-cube, like the line and the square, has a direction associated with it which would be at right angles to the next defined motion. Labeling this unit as positive continues the vector content in accord with the reversal idea upon which it is based. As we have seen above, *each operation represents a change in direction, but of 90 degrees, not a total reversal of 180.* If we go, + 2, +2, -2, in our sequence of operations, we will go to the right first, up second, and back from the "center-plane" third to form another eight unit volume which will be above, to the right, but behind the point of origin. This will be considered a "negative volume" purely by convention as it has one negative sign associated, however, this convention does preserve the vector designation by the conventions observed.

-

As the positive numbers are associated with "positive values," right--as in handedness, upward--toward the Heavens, and forward--"progress." Negative numbers are associated with the reverse, left--"sinister" or left-handed, downward, and backward. The operation, -2 x -2 x -2 , would create a cube, which was to the left of the origin point, below the 'origin line," aka, the "x-axis" and behind the "origin plane," aka, the "xy-plane." Note that the first square formed would be considered a "positive number as it is "minus x minus = plus" but the third operation, adding another direction considered "minus" labels the resulting volume as a "negative number volume."
Summing the above, a signed number represents a line, two signed numbers multiplied together represent a plane and three signed numbers multiplied together represent a volume and, in the order of the multiplication process will determine what plane or volume is described. The operation, +2 x -2, represents, by the conventions used, a square which is to the right and down from the origin. while the

reverse operation, -2 x +2, forms the representation of a square which is to the left and above the origin.

It can be seen that 1 x I x I as counting numbers still represents the original one. Plus-one times Plus-one times Plus one represents one whole, but it is one whole cube, one length to a side, not the original one which would be one line, in this "signed" case.

Similarly, it can be seen that the cube root of eight as a counting number is simply the number two. The "cube root of +8, as a "vector cube" has the "absolute value" of 2 but this two can be either a positive or negative vector depending on which of the genus it belongs to and the order in which it falls in the set. A positive-volume-vector, +8, can be generated by any of four sets of three "signed twos." These sets are as follows: {+2, +2, +2}, {-2, -2, +2}, {-2, +2, -2}, or {+2, -2, -2}. A negative-volume-vector,-8, can be generated by any one of the sequenced-operation sets {-2, -2, -2}, {-2, +2,+2}, (+2, +2,-2} or {+2,-2,+2}. Assigning a "signed-root" to a signed number, is a difficult and tricky business which would actually require a knowledge of the history of the signed number in question! It is no wonder that the mathematicians seem to ignore "odd-number" roots of signed numbers and consider that the square root of minus one is "plus or minus 'i,' an imaginary number, which truly it is, for in the most basic unit of "minus one" we are considering a line vector of a unit length and how does one take a root of a line vector "running backwards?" Actually the root would have absolute dimension of one, either plus or minus, as one would have to be speaking of the "second-order-vector-square" which can be generated by either of the sets, {+1, -1} or {-1, +I}. Mathematicians have no trouble with saying that the square root of +1 is plus or minus one as it is generated by the two sets, {+1, +1} and {-1, -1}, sets within which the internal values appear to be identical to one another. As one can see from previous discussions that the internal elements are not identical but represent different directions of the vector depending on their position in the sequence.

-

The use of the two signs with different meanings of operation, reversal, or direction causes some interesting problems in understanding mathematics.

-

-

The following paper is a follow-up, companion piece, to the above paper, it.too, was posted on the SciScoop Site.

Roots and Directed Numbers

In working with numbers one often works with fractional exponents, "roots." This works well when one is working with "absolute values," unsigned numbers; however, it runs into complications as soon as one starts to operate with signed numbers. The problem probably arises from the inherent fact that absolute number values do not have a directed motion automatically assigned to them; whereas, signed numbers do. A positive number is associated– usually–with a motion upward, to the right, or forward, with a negative number associated with motion downward, to the left, or backward.

 When we take the square root of one, unsigned, we're are talking about a number which multiplied by itself gives the original number, one. When we take the square root of four, we realize that it is the number two, when we place two units down twice we get four. When we are working with signed numbers we have a different situation.

 If we are taking the square root of +4, we are actually asking the question, "What is the directed side length of a square which we consider to have the area 'Positive For' when we operate according to the conventions associated with signed numbers?" By those conventions we can see that both +2 times +2 and -2 times -2 fit this criterion, so we say, quite correctly that the square root of +4 is either +2 or –2, Perhaps we would, however, have been more accurate in saying that there are two sets of square roots to the number, +4, the set, +2,+2, {+2, +2} and the set ,{-2,-2}.

The reason that this last was said will become clear when we discuss the situation for the "square root of -4." Let us analyze this problem as we did above. The question we are asking "What two directed numbers will produce an area which by our conventions of directed numbers will be assigned a value of -4?" This occurs again in two cases, producing two sets, {+2, -2} and {-2,+2} . As these are directed numbers the set, {+2,-2} is not identical to the set {-2, +2} as they represent opposite directions of sequential motion. With the "Positive Area" we find that we can create what we call a positive area by going in a positive direction then turning in another positive direction, or going in a negative direction and then turning in a negative direction again. For a negative area we can start out in a positive direction, then "turn negative" or start in a negative direction and "turn positive." By this analysis, the square root of "Negative One" is not an imaginary number but can be said to be not as in the other case, "Plus or Minus One" implying each "operating" on itself, but "Plus and minus one" the two operating on each other. The concept of imaginary numbers arises because of the ignoring of this fact of the directed action factor inherent to signed numbers.This can, of course, be extended to higher roots. For the cube root of +8, one may write the sets, {+2, +2, +2}, {-2,-2,+2}, {+2, -2, -2}, and   {-2,+2,-2}. Noting four sets that can be considered the "cube root " of +8. A similar group of 4 sets represents the "cube root" of -8. A fourth root would presumably continue the pattern developing, eight sets of 4 units each. This is left to be proven, or disproven, by the reader. The idea of an endless set of imaginary numbers as successive even roots of "Minus One," is an interesting concept; but, by the above analysis, appears to be based on a misunderstanding of the significance of signed numbers. The use of a signed number indicates a motion in a direction and can be considered to define a "dimension."   Posted on SciScoop March 17, 2009,

  Essays: The concept of time There are a number of Essays on the concept of "Time" to be found under this title on Helium.com.The following short essay was rated "22 of 42" at the time it was copied into

Google Drive a couple of years ago. This article reflects a simpler, more restrictive, view of "Time" as compared to, for example, "Minkowski Time" which seems to extend the idea of "Time" to a "Dimension which encompasses all possible sequences stemming from any given starting point. " *(This is the writer's own interpretation of Minkowski Time, and is not necessarily Minkowski's intention.)*

Time is a human construct used to keep track of sequential motions in space transfer, it is sometimes considered as if it were a fourth spatial dimension.Time is usually considered to have three aspects, the past, present and future. These can be considered to correspond to motion of a specific point on a wave front. The past is the total sum of all motions that led to that instantaneous position which we consider the present, and the future is where that wavefront spot may be considered to go in the next stant and all the instants which may follow.

The present is the result of a certain sequence of motions, which we call the past, something which no longer exists, but which is nevertheless a "fixed construction."

To go back into the past, in a physical sense, would require a retrograde repetition of all of these motions, a set of motions which would increase instant by instant in a huge geometric progression. Our "wavefront" would have to move backwards in a perfect reversal of the sequence by which it had previously moved forward.

Even were this possible, one can see, that, since the direction has been reversed, what was "back" is now "forward," the wavefront, while "retracing the past" would actually by "moving into the future." In trying to go into the past, a "time-traveler" would actually be attempting to create a future which was a reversal of the past.

If one considers that it may be possible that long wave fluctuations in the "Matrix" in which we exist creates multiple adjacent universes in which certain sequences may coincide, it might be possible to move from one alternate universe into another which would correspond exactly to some point in one's own past. There is no real indication that there are such alternate universes; and, were they to exist, it is highly unlikely that they would correspond in such way as for there to be possible entry from one to another. However, if one wished to combine Religion, Brane Theory, and ultra-slow vibrations in the Matrix of Existence, maybe "Heaven is Just a Brane Away?" That sounds like a new hymn, or Country-Western Song.

While we are on the subject of mathematics, we might as well take an essay from the files on the most famous mathematical equation of all time. This article, written about 2012, is slightly dated, some of the definitions have been extended; e.g, the definition of "Mass." However, the basic ideas seem valid. (ds, Jan . 2015)

The meaning of mc^2.

In the half decade, or so, since the first articles on this subject were posted on Helium.com, there have been advances in the consideration of what constitutes Mass and Energy, leading to definition of these in terms of Motion, which allow mathematical derivation of the term, "mc^2," in at least three different contexts.

A definition of Mass has arisen: "Mass is a measure of the motion content within a surface, as a sphere, considered as being focused at a point in the center of

the sphere, and measured at a point on that surface."
This is usually measured by comparisons to some
standard.

A working definition of Energy is, as follows: "A
general term for a packet of motion. As usually used it
means motion along a line which results in collisions,
and is usually measured in some manner which is
related to those collisions, This is called "Kinetic
Energy, and is usually considered to be related to
:"Mass" by the equation, $E=mv^2$, where "v" is the
relative velocity of the colliding particle before striking
an immovable object.

This Kinetic Energy definition, leads directly to the
$E=mc^2$ expression. That is, if one considers that two
units having identical "vector velocity components of
'c' collide head-on," stopping to a :"zero" situation
and dissipating the combined motion into the
surroundings.   This apparently is exactly what
happens in the "annihilation" of electron and anti-
electron, who, after spending what would be ''eons'' in
each other's vicinity (*were we to be able to watch
them in a time frame referenced to their own rotation-
inversion-rotation-of-inversion cycle*) finally reach an
orientation
wherein they can do a "Yin-Yang" combination into a
neutral entity, dumping the rotational motion at the
speed of light, which had given each a "net charge."

Another way to obtain the expression, $E=mc^2$, is to evaluate the "Mass" that would be associated with the wave front of a light wave. This involves the use of three fundamental equations. One is the definition of the speed of light as a frequency times a wave length. The fact that the wavelength of light can be considered as the radius of a sphere at the wave front, can be fed into the "Lost Equation," *mass times radius equals Planck's Constant divided by the Speed of Light.;*" by converting it to a frequency. The result is an equation, *$m=(hu)/c^2$,* as the definition of the effective mass of a light wave front. *Since by Planck's Equation, "hu," Planck's Constant times frequency, is "Kinetic Energy,"* we have a formula which can be rearranged to $E=mc^2$.

This derivation points out the interesting fact that the "measured mass" of any object, may well be. in essence, the Wave Front Kinetic Energy which is the result of the collisions of its internal parts in such a way as to cause a true circular wave front having an effective mass which is convertible to "Kinetic Energy"--that is, if the internal units were allowed to escape..

This suggests that what we measure as "Mass--"and as "Mass convertible to Energy" by $E=mc^2$--is but a small fraction of the total motion content of any unit. That is, the part of that internal motion which is due to internal collisions, i.e. "heat," and *not due to the other vibrational-rotational "point-centric" motions which may be characteristic of the internal components. That is the inherent internal motions of the electron/positron, proton/antiproton entities that constitute the basic units of "Matter."*

There is still another way to derive the expression, "mc^2, which is a bit more theoretical and leaves a bit of mystery yet connected to the expression. The derivation goes like this:

Momentum, mv, p, is considered as the rate of change of Energy with respect to Time, the *"derivative" in Calculus terms.* If we do the "Integration" with respect to time considering mass constant, we get the familiar expression, *$E=(mv^2)/2$.* If we decide that mass could change at some constant velocity, and do that integration, we get an unfamiliar form, say, E', of the form *$E'=(vm^2)/2$.* If we just integrate "p," we get a third expression, E", *$E''=(m^2v^2)/2$ when we substitute, mv, back in for "p."* When we evaluate this last expression at "v=c" to obtain E"=$(m^2c^2/2$ and "back differentiate," considering "c" constant, we get our old friend, "mc^2," back; but, this time, with the meaning that it is the rate of change of mass with respect to time at a constant velocity of "c."

Well. there you are. We have. three evaluations of the possible meaning of "mc^2," none of which are exactly the same as the usually accepted meaning of "mc^2" as a conversion unit for all of what would constitute the "Mass" of an object into "Energy." However, these meanings are quite dependent on the definitions of "Mass" and "Energy" as being different aspects of motion within a substance of undefined extent and undefined basic unit, rather than Mass and Energy being rather undefined, interconvertible "somethings."

(Comment: Jan. 2015, It is interesting that Einstein never seems to have given an explanation or derivation for the Equation; however, it does appear to derive naturally from the photoelectric effect for which Einstein received a Nobel Prize.

The following article attempts to explain the basis of using the Balance Equation of Nature to predict from limited data.

THE CONGRUENT PARALLELOGRAM THEOREM AND A PERCEPTUAL UNIVERSE

The "Congruent Parallelogram Theorem" applied to two constants of nature, Planck's Constant, "h," and the Speed of Light, "c," and the Torque Constant of Nature which arises as their ratio, "h/c," produces a description of a "Perceptual Universe," in which "h," and "c," are valid descriptors.

The right and left halves of the mathematical relationship, $Ax \times By = K = Bx \times Ay$, can be said to describe, "congruent parallelograms."
This is independent of the dimension units attached to the variables, "x," and "y." This relationship, which may be called, "The Congruent Parallelogram Theorem," is widely applicable in the field of physics where it appears in a number of guises. It is in the Law of Levers, the simplest expression of the Conservation of Momentum and of the Law of Forces, "For each and every force there is an equal and opposite force."

In general, it may be stated, " If there be a be a constant. "K." which may be analyzed as the product of two factors, the theorem is applicable in analyzing implications of that constant." Applying this to Planck's Constant, "h," the Speed of Light, "c," and the Torque Constant of Nature, which is their ratio, "h/c," produces a description of a "Perceptual Universe, " in which these constants are valid.

-

Writing the definition, " The Speed of Light is the combination of the Frequency and Wavelength, in the Theorem form,
Aw x Bf = c = Bf x Aw, where "f," is a frequency unit definition, "w," a wavelength unit definition and "A." and "B" are the absolute values associated, We see that for any associated pair of wavelength and frequency there will exist an "exactly" congruent set which can be found by reversing the associated absolute values. Another very interesting observation is that there will be an "instant" or "set of instances" wherein the absolute values of the frequency and the wavelength are exactly the same, i.e., $f = w = c^{0.5}$. This interesting relationship will be explored further at another point.

-

An important set, which may be considered to define the "Upper and Lower Cut-off Communication Frequencies for a Perceptual Universe" can be found in the following manner. The starting point is the Planck formula for electromagnetic radiation--thought to be the fastest communication method known-- Energy equals Planck's Constant times Frequency, $E = h \times f$. Realizing that the totality of motions involved in the entire "Energy" of this Perceptual Universe may be represented as a Unity, "One," we write, "$1 = h \times f$," and see that the corresponding frequency would be "$1/f$." the "reciprocal of Planck's Constant." the corresponding wavelength, found from, $c = f \times w$, is "ch."

 This very high frequency and short wave length would be the "High Frequency Cut-off." The reversal, " frequency, ch," and "wavelength, l/h," would define the "Low Frequency Cut-off." (In the cgs system of units the high frequency cutoff would be at about $1.5 \times 10^{+26}$ cycles per second at a wavelength of about $1,97 \times 10^{-16}$ cm. and the low-frequency cut-off at 1.97×10^{-16} cps. and a wavelength of $1.5 \times 10^{+26}$ cm.

-
Additional information arises from the "Torque Constant, h/c." This expression is found to be a torque, "mass times radius," by equating Planck's Constant to its definition a s an angular momentum, "mass times radius times velocity. to get the expression,
$h = m \times r \times v$ " where "m," is a "mass," moving at a "tangential velocity, v, " at a distance, "r." from some defined center. Into this expression is inserted "c--" in

its role as an average velocity in any given direction--for "v." This produces the equation,
h = m x r x, which, rearranged into the Theorem form, m x r = h/c = r x m , can be seen to be an equation which defines a "family" of oscillators of torque constant, h/c, with an average values where m = r = $(h/c)^{0.5}$. This last relationship, $(h/c)^{0.5}$, would define a central sphere or circle through which all of the oscillators of this family of the set,
{m x r = h/c = r x m}, would invert. This radius value is about 4.7×10^{-19} cm. The "mass," measured for any oscillator at this distance, would be about 4.7×10^{-19} grams. (Our postulated "Universe," has some interesting coincidences in the areas of Quantum Mechanics and String Theory. Quantum Mechanics is said to fail at below 10^{-18} cm., essentially this inversion radius. The strings of String Theory are said to vanish into a "Hole," at 10^{-18} cm. This "Hole" would have essentially the diameter of the inversion circle/sphere that appears in the above reasoning.)
If it be assumed that a wavelength associated with this average unit is the circumference of the unit, i. e.. $2 Pi (h/c)^{0.5}$, a possible average oscillator frequency for this "Universe," would be "$c/[2 Pi(h/c)^{0.5}]$" which is equal to $c^{1.5}/ 2 Pi (h)^{0.5}$. As "Time" can be seen as the reciprocal of frequency, a basic time unit can be defined as $2 Pi \times h^{0.5} /c^{1.5}$

One can assume a wavelength associated with the radius, to obtain a simpler set of expressions, $c^{1.5}/h^{0.5}$ and $h^{0.5}/c^{1.5}$ This type of assumption would agree with the situation which arises when one checks out the electron and proton as members of the set {m x r = h/c}. when rest mass is considered as one limit, the radius is found to correspond to the "Compton Wavelength" of the electron or proton.

-

Going back to the situation of $f = w = c^{0.5}$, one can check out what will result if one looks at an oscillator based on this situation, assuming that "w" can be taken to be the equivalent to "r" for a limit of an associated oscillator of the "h/c set." If $w = r = c^{0.5}$, then the corresponding "m" equals $(h/c)/c^{0.5} = h/c^{1.5}$, and the reversed limits are $r = h/c^{1.5}$ and $m = c^{0.5}$ That is, this oscillator would be a sphere containing an internal sphere, the outer sphere having the radius of $c^{0.5}$ and the inner sphere would have the radius of $h/c^{1.5}$.

-

In every case we are assuming that we are operating with the absolute values of "c" and "h" in whatever consistent set of units we care to use and that the square root values are also used as absolute values.

-

In summary, applying the Congruent Parallelogram

Theorem to two basic laws of nature and their ratio has resulted in the definition of some of the characteristics of a postulated "Perceptual Universe" wherein communication would be controlled by these units. There has been defined a high and low frequency cut-off and an average size and mass for a family of oscillators that would be presumed to operate in this postulated Universe.

This postulated "Perceptual Universe" can be extended as a possible model for "Our Reality," by evaluating the electron and proton as possible oscillators of this set and bringing in additional information. To an extent this is essentially what has been done in "pages," published on the Internet site, URL, http://groups.google/group/oscillatorsubstance-theory. On that site the subjects of "Virtual Electrons" and "Super Symmetry" have not been addressed. It is possible that both of these concepts would have pertinence to our "Perceptual Universe" as developed here, through the application of the Congruent Parallelogram Theorem to frequencies associated with the electron and proton. However, this is beyond the scope of this paper and should be addressed as a separate topic.

A Critique of Einsteinian Relativity

There is so much mystique, and so much elegant mathematics attached to Dr. Einstein's ideas that for one to even hint in the presence of most physical scientists that there is a flaw in Relativity is like standing on a street in Vatican City and averring, "Jesus was a Zealot leader." Saying that the "no-speed-faster-than-light" dictum has a definite limitation in application is tantamount to heresy. Einstein's Theory of Relativity--now called the Special Theory of Relativity, after he generalized his ideas from moving "frames of reference" to accelerated frames of reference--has been treated as "fact" by the scientific establishment. It is long since time that his ideas were reexamined to determine to what extent the conclusions are indeed fact, and to what extent they have been misapplied.

Let us look at statements of Dr. Einstein's two basic premises in a couple of different ways and see what happens.

Premise 1. There is for every observer a frame of reference in which the normal laws of physics hold true. (This is, I am sure, not an exact translation of his original wording, but should carry the original intent.)

Premise 2. There is no possible speed greater than the speed of light. Nothing can move faster than the speed of light. (Again, undoubtedly not an exact translation; but it is the way that the second premise is generally understood.)Now, looking from a different viewpoint, we can write two statements which make essentially the same point.

Statement 1. All information coming to a receiver from a transmitter which is not in motion with respect to the receiver will not be distorted during transmission. (Things that are in the same frame of reference are not moving with relation to one another.)

Statement 2. No information can be transmitted at a velocity greater than the speed of light. No information nor energy can be transmitted at beyond the speed of light. (The fastest carrier wave of information and energy of which we have any knowledge is electromagnetic radiation.)

What we have done in the above is to restate Dr. Einstein's premises in the form of statements having to do with the transfer of information. Having done this, we can see clearly that the ideas are clearly fact as related to information transfer.

The next stage is wherein the mathematical equations relating "rest values" --that is, the values of any

measurement made wherein objects are not moving with relation to one another--and, "relativistic values"--the values that measurements will apparently have when they are transmitted between transmitter/receiver pairs that are in motion with respect to one another. These equations accurately portray the distortion that will occur in such cases; but, can be misinterpreted and misapplied. These equations are all of the following form:.
$A^*=A/(1-v^2/c^2)^{0.5}$, where, A* is the "Relativistic Value." A is the "rest" value. "v" is the relative velocity between transmitter and receiver, and "c" is a constant, the Speed of Light in a vacuum. These equations become meaningless when "v" is equal to, or greater than,. "c." The "Relativistic Value" becomes "Infinite."

 The term, "Infinite," is usually interpreted, erroneously, as, "Totally beyond all measurement." What it should be interpreted as is, "Not measurable by the assumptions were are making, or, the tools we are using." The first interpretation leads to such ideas as, "The mass of an object moving at the speed of light is 'Infinite.'" The more correct statement would be, "The mass of an object moving with respect to an observer cannot be measured accurately by any means of which we are aware. " Another correct statement would be. " If an attempt is made to accelerate an object to beyond the speed of light with energy supplied from an external source, the effect of attempting to transmit energy at the speed of light will make the object appear to have 'infinite mass.'" The mass of the object will not have been changed, its "apparent mass" is changed by the problems of information/energy transmission.

Since equations of the form cited above can be generalized by inserting other "carrier wave" velocities in place of the speed of light, the Einsteinian Relativity ideas can be applied to an unlimited number of what could be called "Perceptual Universes" which could be defined by a particular viewpoint and a particular maximum velocity of information transfer. One could use the basic ideas in Sonar Research, in research having to do with nerve impulses, and possibly in many other ways. Einstein's work is totally valid as long as we are talking about information/energy transfer.

There are, however, several misinterpretations which seem to be pervasive in the scientific community. One is that "relativistic values" are taken as having total reality. Relativistic effects at a receiver may be the same as if they were "real." If we consider the "real" facts as measured closest to the origin, we can see that "relativistic values" are distortions.

Another pervasive misuse is the application of the

no-speed-greater-than-light dictum to the relative motion of freely moving bodies, and to the maximum velocity along a predetermined vector which a moving object might attain. A few moments of reflection should show any thinking person that application of the dictum to these cases is almost surely arrant nonsense. However, these two misinterpretations seem to be generally accepted.

Dr. Einstein, himself, does not seem to have realized that his work more properly belonged on the area of information Theory than in physics. The extension of "Special Relativity" to accelerated systems resulting in General Relativity, has the same problems related to the misuse of the ideas to systems/situations wherein information/energy transfer is either unnecessary or impossible.

[The Space/Time model can, also, be related to "Information Theory." If one wishes to identify exactly a happening, one must tell "where" and "when." In mathematical terms one must identify the happening in three dimensions of space and one dimension of measured sequence. If one is unsure of a location in three-dimensional space, one can "triangulate" it from three known points. Having done this, one would yet have to locate this point with reference to a sequence. Now if you don't have any reference points to start with you can do what Einstein apparently did,

he simply set up three dimensions (vectors) for each of his unknown reference points, then added another dimension to each, converting his "3-D vectors to "Tensors." Having done so, he said that he could describe Space/Time with nine tensors. (More recently, Hawking, et al, in String Theory, talk about a ten-dimensional universe. Presumably, this "10-D-Universe" is based on the same idea, as nine "tensors" would add up to ten dimensions.]

In summary: Einstein's work seems valid when applied to cases wherein Information or Energy is necessarily transmitted/received. It has no apparent validity in cases wherein such transfer cannot take place.

The following Little article which was published on a closed Internet site by the writer, illustrates a rather dim opinion of certain types of very popular physics research.

PHYSICS BEYOND THE STANDARD MODEL?

The question seems to have been asked somewhere, "Is there Physics beyond the Standard Model?"
the answer has to be a resounding "YES!" The Standard Model is a "Crock." based on misunderstood data and toggle up after toggle up.

The "Discovery of the Higgs Boson" this Summer was a wonderful exercise in wishful thinking, for which the scientists who made the report will probably get a Nobel Prize.

After all, they had to find what they were looking for or admit that the whole thing was nonsense. They found indications of something. Although it wasn't very near where they thought the signal would be. The finding of *something* proved to them that they had found what they sought.

I call this kind of research,"Pelican Shooting," from an old story about a novice goose hunter who blazed away at the first flock of birds that went over, then wondered why the dog didn't go to retrieve. An old hunter finally advised him, "You darn fool, you shot a pelican."

Physicists seem to love this kind of research:
Get an idea. Spend lots of money, and, if they find anything, claim that it is what they are looking for.

The following was excerpted from EGO OUT, the Blog of Dr. Peter Gluck, concerning the myth of "Antimatter Annihilation."

Wednesday, October 15, 2014

ABOUT "A SCIENTIFIC FACT, GRAVEN IN STONE"

It is known to every physical scientist that it is an absolute Fact, carved forever in stone, that Matter and Antimatter annihilate on contact.
This supposed "Fact" is totally refuted by logic stemming from two other quite incontrovertible facts.

1. Physical objects which are mirror images, or very close to mirror images, when placed in a particular orientation to one another combine into a far more symmetric unit. They do not "annihilate," totally destroying one another, they simply combine.
2. Matter and Antimatter, by the very definitions, may be considered as perfect, or, very near perfect, mirror images.
Conclusion: Matter and Antimatter units, if they can be brought to an exact orientation will combine to form a combination unit. This is what might be called a"Ying-Yang."

As we now have one unit rather than two, there will be but half of the total motion disturbance that there was before. "Energy will be emitted."
Here are some examples with possibly logical interpretations.

1. Negatron and Positron. (Electron and Anti-electron.) This is the "Annihilation" which has been often observed, misinterpreted, and over-generalized. It can be diagrammed as follows:

1, $e^- + e^+ ---> e^0$, "Zerotron formation." This unit has never been observed, probably because it was never suspected, yet it may be the "Ubiquitous Parent of All Matter."

2. Proton and "Conton," diagrammed similarly produces what could by analogy be called a "Zeroton." This unit, as with the "Zerotron," has never been suspected. However, it may be a known particle, the "Tau."

3. Hydrogen and Antihydrogen combine. The product? Deuterium.

4. Deuterium and Anti-Deuterium combine. The product? Helium Four.

5. Tritium (H3) and Anti-Tritium combine. The product? Lithium Six.

Although we could continue to write other hypothetical examples, this will not be done. the ones given above are for simple enough units that the anti mers have a finite probability of meeting in the right orientation to combine. For larger units, there is too much complexity for there to be any logical expectation that they would combine, There might be a finite, but very small, possibility for units which form molecular dimers as do all of the cases cited above.

ALTERNATIVES TO THE " NEUTRON-PROTON NUCLEUS" MODEL

THE DUAL-WAY ATOM MODEL
There is an interesting way to combine two different models of the atom to predict how the particular atom, or, perhaps more explicitly the particular " Iso-set of electrons and protons in atom form" will change.
:
(Both of these models derive from the "Oscillators-in-a-Substance Model of Existence" which,by its immediate suggestion of a single Force affecting two different sizes of vortex oscillators, gives a different model of the "nuclear atom," as a unit which which has a "nucleus," not because of some strange forces

between protons and neutrons; but , simply because the comparatively small inversion range of the proton/antiproton entities allows them to pack far more closely than the far wider ranging electron/positron units.}

Dual Set/Subset Model

The first model is based on the familiar "Set and Subset" modeling used in determining the ""electron structure" of an atom. The standard outer electron structure of the atom is taken to be a set that characterizes "electron-proton" coordination and is called here, "Outer electrons."

There is, however, another parallel set--developed in this modelling-- which is perhaps best considered as positron-antiproton coordination. However, as almost none of us is used to the idea of what we know as "Matter" being made up of both "Matter" and "Antimatter" units, we shall call this set, "Inner Electrons." This set is the "neutrons in the nucleus" set by conventional nomenclature.

In both cases, filled and half-filled shells are considered sets which will be favored configurations.

We shall look at a simple set of three Isomers. The

set of Li8, Be8 and B8 , (The mass number eight isotopes of Lithium, Beryllium and Boron.)

The electronic outer structures are, respectively 1s2, 2s1; 1s2, 2s2; and 1s2, 2s2, 2p1.
The electronic. inner structures are, respectively 1s2, 2s2, 2p1; 1s2, 2s2 and 1s2, 2s1
Looking at these, one can see that the middle unit consists of duplicate sets, both of which are filled subshells and each of the others can be changed to it by moving one electron, inner to outer in the first case, outer to inner in the second case. In truth this happens Both transforms take place. The other two are unstable with respect to the Be8 unit, to which both convert.
A second model also arises, which ; we may call the "Deuterium, Tritium, Helium Three (D, T, He3) Model"

Every atom, beyond the Tritium/Helium- 3 isomer-pair can be considered as if it were made up of basic units of Deuterium ,Tritium and/or Helium 3.

Looking at the set we just examined above, we may consider Lithium 8 as a fusion of two Tritium atoms and a Deuterium; Beryllium 8 as a Tritium, a Helium Three and a Deuterium: and Boron 8 as a combination of two Helium 3 Atoms.,and a Deuterium. The central unit (Be6) can be formed by a switch of one electron within either of the other two units. In addition, since T and He3 can switch one to the other the central unit might be considered as "Resonance Stabilized" as, at any one instant, it appears to be impossible to tell which is the T unit and which the He3 unit.

Extension to possible involvement of ions in radioactive decay.

The Be8 unit above would be expected to be stable: however, it is known to "decay" quite quickly to produce two Helium atoms. The decay may be, actually, of the Be++, dication.   The Be8 unit may also be written as the set {He4, He4] as a "resonance form" with the {T, He3, D} set.That would seem to make the Be8 more stable. The di-cation can be written as (He4+, He4+) or as (He4, He4++) . It appears that this last form is the one which collapses into He4 and an isomeric form of He4++, the "Alpha particle."

Switching to the other model, we see that He4, is 1s2 both inner and outer, whereas the Alpha is an interesting situation as we really have no way of knowing truly whether the two missing charges are outer or inner or even a "split" situation of two "half-filled" shells, "1s1," with no way to tell which be inner or outer set! Clearly the set, {He, Alpha} represents units which will have different shapes and forms, and be capable of independent existence. Be8 is the first "Alpha Emitter." It is, also, at the same time, the first example of spontaneous fission, and illustration that fission is not symmetric.

In future essays, there is intended to be a look at other sets. Ideally, of course, one would look at isomer sets up to the ones including the Transuranium elements. This would be a massive undertaking.

Two cases that should be interesting is to try to determine why Bi83/209 is the last "stable" unit , and why none of the isotopes of Technetium are long term stable.

The following two papers are a result of the realization that it is possible to consider all known naturally occurring isotopes as being made up of three basic units, Tritium, Helium Three and Deuterium, that is skipping the fact that all could be considered to be have formed originally from Hydrogen One in some way.

The result is an interesting new way to look at Elements as this type of coding emphasizes different "Families" of isotopes from the usual listing as Elements.

BASIC UNIT CODING OF ATOMIC NUCLEI
An Attempt at a Very Basic Primer

Atoms, which are considered the building blocks of everything which we call matter-- anything that can be weighed--are usually considered to consist of a very tiny central region called the "nucleus," and a much larger outer region having no specific name, which is filled with "electrons."

The inner region is said to have almost all of the "Mass" of the atom. Mass is what we measure by comparison to some "standard mass," a unit of which we know the "weight." .This tiny region is said also to have a "positive charge," which is equal to the total "negative charge" of the "electrons," the much larger particles which fill up most of the space of the atom.

(To be more rigorously correct, theoretical considerations suggest that the electron has a far wider range of oscillation than does the proton--the basic positive-charged entity-- and is, therefore, actually both far larger and far smaller than the proton: however, the "far smaller" aspect is rarely noted, and will not be further discussed at this point. .)

All of the atoms which occur in Nature, have nuclei that may be considered as made up of four smaller, basal nuclei.These are, as follows:

A nucleus having three "Units of Mass" and two "Units of Charge,

A nucleus having three Units of Mass and One Unit of Charge,

A nucleus having two Units of Mass" and One Unit of Charge,

A nucleus having one unit of Mass and One Unit of Charge.

This last listed, basic unit is called the "Proton."

 Since almost all the known nuclei of atoms larger than a proton may be considered as made up of some combination of the other three units which are listed above. we shall not say much about the proton for this short discussion.

Use of the above should describe atoms in a way that may be very useful in computers. by using a set of four columns that correspond to the above four basic units.

The form of table used by this writer to do this is to make the first column--from the Viewer's Left-- represent the number of mass-three-charge-two nuclei, These would be called, in current literature, "Helium 3 Di-cations. The second column is the number of the mass-three-charge-one nuclei. The "Triton Cation". The third column is the number of mass-two-charge-one nuclei, known in the literature as the Deuterium Cation. The last column would be the "One to One," very basic situation called the Proton, or the Hydrogen cation.

(This is a comment explaining some terminology. If the reader is familiar with fundamental chemical terminology, this

parenthetical section can be skipped.

 A unit which has a 'positive charge" is called a "cation."

 Visualize a cat with a positive (+) eye?

That is what I did to remember the term.

The mono-cation of "Mass One" is called a proton.

 The term for a unit having a negative, or minus, (-), charge is called an "anion." Yes, that is what it is, "An ion is negative, if it isn't catty...."

 One can use any bad pun or silly picture to remember a term!

 The very simplest, "almost massless--in usual thinking," anion, is the "Electron" or "Negatron."

 The simplest, fundamental cation, is the "Electron's Mirror Image," the "Positron" or "Anti-electron."

 For our discussion of the composition of nuclei in the above coding, we shall ignore these two very fundamental units.)

Resuming the original discussion:

Our fundamental nuclear units we would write down as below, giving the name, then what it would look like in our four column coding:

Our first unit, of Mass One, the "Proton," would code,   O,O, 0, 1

The second, called the "Deuteron," would code,   0, 0, 1, 0

The third unit, called the "Triton," comes in as   0, I , 0, 0

and that first one from the Left, "Helium 3 Cation," is   1, 0, 0, 0.

The next unit, "The Helium 4 Di-cation," we code as "2 2's,"   0; 0, 2, 0. (An isomeric form of this, the Alpha Particle, would be given the same coding. (Isomers have the same numbers of the same kinds of basic units, but different geometric forms.) When we get to much larger units, it becomes a bit of a mathematical

problem to figure out the coding. We adopt the convention of "pinning both the masses and the charges" as far to the left in our coding columns as we can. As it turns out, we will almost never have a number in the last column to the right, except zero, and "One." The latter case will occur very rarely. The next column, of mass 2, will have only the numbers Zero, One, or Two.

 The challenge comes in determining what numbers to assign to the units of "Mass Three."

To see how this works, we will jump to something which all of us are interested in, *GOLD*.

. There is only one unit found in nature which is called "Gold," and its nucleus is said to have a "Mass number of 197 and a charge number of 79." This nuclear charge number is called the Atomic Number and the Mass Number is usually called the Atomic Weight.

To determine what numbers to assign, we first check to see how many units of three we would have, if all of the "three" units were to have a charge of one, This is easy to do. We divide by three. One hundred ninety seven, divided by three, is 65, with a remainder of two.

 So we have sixty five units of mass three, and a unit of mass two. We can immediately place a "one" in the third column from the Left, as we have found a "Unit of Mass 2", which we know will have a charge of one. That remainder told us what we were first testing for.

The filling of the easiest column to fill, the "2-1", Deuteron, column.

 Now how do we find what the other numbers are?

We have accounted for one of the 79 charges. We have 78 yet to account for. We have 65 places that will have either one or two units of charge. Clearly then, we can find how many charges have to be doubled up by putting one on each of the 65 and doubling up the remaining 13 on units that we already placed one on. We have determined the number of units of Mass 3, charge 2, as being "13." We place that in our first column to the right. At this point we have the situation, "13, ?, 1, 0. "

How to determine the last column, the "Mass three, charge one" column?

Do another subtraction.

We started with 65 units of two with one charge on them and removed

thirteen by putting two charges on them. What we have left would be 65 minus 13 equals 52 units to fill in for our question mark.

Our final "name" for natural Gold, will be "13, 52, 1, 0." Which we may shorten to "Thirteen, fifty two and one."

(Now, Dear Reader, you can "Snow Job" your friends by saying, "I'll take all the 'Thirteen, fifty two and one' that I can get." They'll think that you are talking about, well you know..

My friend, Charles William Johnson of Earth/matriX, points out that 13 and 52 are Maya Calendar Numbers.

On a bit of reflection, it is seen to be an interesting coincidence that the above coding for Gold, which is often associated with the Sun, would be easily associated with a logical calendar of 13 months, 52 weeks, and one day, 365 days.

Probably pure coincidence, but interesting.)

Had we been talking about another "isotope," of Gold--a form not found in nature--having a mass number of 196, in dividing it out, we would have found the same 65 units of three, with, this time, a remainder of one.

("Isotopes of an Element" are units having the same "Charge Units" on the nuclei but different "Mass Unit" values.)

We do not want to put anything in the one's column if we don't have to; therefore, we will, instead, say that we have a set of 64 units and two sets of two. We place a "2," in the "Mass 2 column to indicate the two units of mass; and, see that we now have 77 charges to account for. 77-64 = 13. Again, we have 13 mass-three- charge-two units, Now since , 64-13 = 51, this form of "Radioactive Gold," has a final coding of, "13, 51, 2, 0. "

To check our work we work backwards:
2x 13, + 51,+ 2 = 79 and 3 x (l3 + 51) + (2 x 2) = 196. Geez, we did get it right!

(Doing this coding would either give one's fingers a workout on the hand calculator or

force us to remember our early grade school math. which probably isn't taught at the present.)

Had we looked at Gold 198, (Au 198) we would have seen that the division was even at 66. The coding is a "snap." The "last two columns" are zeros. 79 -66 =13. again. and the coding then in 13, (66-13 = 53) , 0, 0. "13, 53, 0, 0."

(It may be noted that when that column which can have a value of Zero, One, or Two contains a "One." the atomic nucleus is likely to be more stable than its neighbors at "Zero," and "Two," units of Deuterons. This seems to be a quite general case which can be rationalized by ideas from a model called the "Oscillators-in-a-Substance Model,'

however, that discussion we can put off to another date.)

It is hoped that this discussion of the coding of three isotopes of Gold has outlined the general procedure that can be used to code any isotope of any element.

This little essay is a "Prequel" to other closely related essays. It is hoped that they can be combined, somehow, into a coherent whole. understandable whole.

The following is a previous article, one of those mentioned above. Hopefully, the two articles together will furnish enough information for others to be able to pick up on the ideas, and, perhaps, find them of utility

NATURALLY OCCURRING ISOTOPES BY THE "FOUR FACTOR CODING"

This article was written before the article on coding the nuclei, and considers the coding of the entire atom by the same type of units, but considering the entire atom as a unit. The coding; however, works out the same coding by Helium 3, Tritium, Deuterium and Hydrogen 1, for the whole atom as we obtain coding by the respective cations for the nuclei in the previous article where three forms of Gold were used as examples.

It is still a mass and charge coding, by atomic mass and atomic number, exactly the same as before.

The four factor coding is a way of coding the

naturally occurring isotopes, and most of the known isotopes, as being made up of the units Helium 3, Tritium, Deuterium and Hydrogen. With each of these units supplying to the "Atomic Number Pool." A number of units equal to the "number of outer valence electrons. That is. in the order listed above, respectively, 2, l, 1, and 1; and, to the "Atomic Mass Pool," an appropriate number of mass units, that is, respectively, 3, 3, 2, and 1.

How this works will become clear as units are coded.

The neutron would actually take on a special five coding, as in some very rare and short-lived units, either it, or something resembling it, seems to add a mass unit without adding an atomic number unit, so the neutron codes as, respectively, as above:

0 0.0.0 /1."With the "slash one,' indicating its different function from the other three units.

Hydrogen 1, H1, codes, 0 0 0 1;

Hydrogen 2, H2, or Deuteron, codes, 0 0 1 0,

Hydrogen 3, H3 or tritium atom, is   0 1 0 0,

and the Helium 3 unit is   1 0 0 0.

Lithium 3, mass 3, is   0 0 0 3.

(With no "embedded electron." this is a very short-lived radioisotope! There might be a great deal of skepticism as to its actual existence.)

Before going any farther, perhaps the question, "Why do this" should be partially answered.

One justification for this kind of coding, which would fit very well into computer work is that many of the transforms of nuclear chemistry may be explained by interactions within an atom.

of the four above units, in fact, many can be explained through changes within .the "Mass 3 Set" of Tritium and Helium 3, which are

interconvertible through a common intermediate. a "(T/He)+." In verbiage, this be the "Common-Cation of Helium 3 and Tritium. "
–

For the rest of this little paper, there will be coded naturally occurring isotopes of elements, with some interspersed comments.

Helium 3 is one of our basic units, coded

above as, 1 0 0 0. It is

coded "first in line," as it seems to be, to some extent, a basic "control unit."

Helium 4, mass 4, valence electrons, 2, 0 0 2 0   Helium 5 , mass 5, valence electrons 2. 0 1 1 0.

(Helium 5 is a short-lived unit, which is included here because of its possible importance as an intermediate in nature.)…

Lithium 6, mass 6, valence electrons 3,   1 1 0 0;

Lithium 7, mass 7, outer elecrons 3 0 1 2 0

Beryllium 8,   1 1,1,0.

(This isotope, Be8, short-lived, important in theory, should be "stable," as a neutral unit. It probably collapses from the easily formed Be++ cation to He4 and an "Alpha Particle." This latter cation is commonly considered to be "the nucleus of a He4 unit,") because it contain the units of 6/8 if a He4 unit.

The remainder of the isotopes to be considered ill be columnated to give an indication of some of the interesting patterns which develop, without pointing them out.

Beryllium 9,   1 2 0 0

Boron 10   1 1 2 0   B 11   1 2 1 0.

Carbon 12   2 2 0 0 .   C13   1 2 2 0.

Nitrogen 14   2 2 1 0   .N15   2 3 0 0.

Oxygen 16   2 2 2 0   O17   2 3 1 0. O18 3 3 0 0.

Fluorine 19   2 3 2 0.

Neon is a 'Perfect, Inert Gas," at Atomic #10 "

Neon 20,   3 3 1 0   Ne21   3 4 0 0. Ne 22 2 4 2 0

Sodium 23   3 4 1 0.

Magnesium 24 4 4 0 0. Mg25   3 4 2 0   Mg 26 3 5 1 0.

Aluminum 27   4 5 0 0.

Silicon 28   4 4 2 0. Si29   4 5 1 0 Si 30 4 6 0 0

Phosphorus 3l 4 5 2 0

Sulfur 32   5 5 1 0.   S33   5 6 0 0   S34 4 6 2 0

Sulfur 36   4 8 0 0.

Chlorine 35   5 6 1 0.   Cl 37   4 7 2 0.

Argon is a "Perfect Inert Gas, at Atomic #18 ".

Argon 36   6 6 0 0.   Ar38   5 7 1 0 Ar40 4 8 2 0.

Potassium 39 6 7 0 0.   K40   5 7 2 0. K41 5 8 1 0

Note: Potassium takes its symbol "K" from the German. "Kalium."

Calcium 40 is a long-lived, radioactive, naturally

occurring unit. It is part of a naturally occurring triad with Ar 40 and K40.Note that all three have the notation , "2 0" in the two last columns. Note also the diad of Argon 36 at 6 6 0 0 and Sulfur 36 at 4 8 0 0, the third member of the triad would be Chlorine 36 at 5 7 0 0 which has too short a half life to be naturally occurring. .

Calcium 40   6 6 2 0   Ca42 6 8 0 0 Ca43 5 8 2 0

Calcium 44   5 9 1 0.

Calcium 46 is a long-lived radioactive isotope, naturally occurring at 4 10 2 0, and, so.also is Ca 48 at 4 12 0 0.

This table, which, in its current longer form runs through the isotopes of Krypton and should someday be extended and elaborated to a book-length document,, is being truncated at this point.(25/3/2015)

Some closing comments as we try to end this at a reasonably readable length.

It may be well to note some implications of this material. which may not be clear.

1. It appears that the Universe within which we live could be considered as a young, living organism which has an "Oscillator Heart," or possibly, a

trio of "Oscillator Hearts," which provide the motion to "keep things going " The idea is that there be a basic high frequency, control oscillator converting a paleo-matter, a basic substance, into a long-term stable proto-matter, possibly what the writer calls the Zerotron. This unit, through shock-wave distortion, is converted to the neutron, from whence all "true Matter" ultimately comes, is an idea which provides a schema, which can be seen to give a rational explanation for some mysteries, including "Black Matter" and "Dark Energy."

2. The "Basic Quantum" of "h/c" and its multitudinous possible forms, with two stable forms at the electron-positron, and proton-"conton " level, gives indication of a "paleo-matter" having a basic motion content as an amount of motion/work/action necessary to rotate one unit once around.

3. The velocity of light provides a number which indicates the amount of these units of a certain size. which tend to interact in information transfer during an arbitrarily designated external cycle e.g. 3×10^{10} units of a centimeter radius per second. It would appear that the protomatter units may be considered as of different sizes depending upon the measurement scales utilized.

That is, since the basic unit, h/c. is about 2x 10^{-37} gram-centimeters the unit of action could be of one gram at 4,7 x 10^{-37} cm. or of 4.7 x10^{-37} gram at one centimeter. .

In any case, Our Universe, in this model, is a constantly changing, constantly expanding unit, most of which is actually still grouped around and expanding outward from the region occupied by the tiny control oscillator or oscillators.

2. The Oscillators-in a-Substance Modelling has little or no effect on many areas of physics and chemistry including basic mechanics, thermodynamic basic electromagnetic theory and kinetics.

However, there are very strong implications in the area of sub-atomic chemistry as well as in basic theory. The model should be general enough that there be possibility to use it to show that while "The twain of SpaceTime and Quantum Mechanics shall never meet," the reason for this statement is that the two older approaches, both actually stemming from the same basic experimental data as this work, are, in one case, SpaceTime, a type of "Substance Model" without the oscillators, and in the other case, o QM. a wave motion approach, closely related to oscillators, which does not have the underlying substance idea. Both appear to be , also, flawed by being signed-number, positive-field approaches. It may be noted, also, that Space and Time are somewhat differently formulated in their basic conceptions, and the hybridization into SpaceTime , while valid as a mathematical model, may have some conflicts in Reality. Space is measured in basic philosophy by placing material objects in ilines Things that can be seen or felt are the the basis of measurements. A "Foot" is based on the length of a human foot. The "Yard" is essentially the length of a human arm. The inch is a "Puglar" in Spanish. The word means "Thumb." It is about the width of a human thumb. Physical objects are used as the basis of measurement standards used to determine physical sizes or physical locations on a line, in a plane or within a volume.

Time is a construct having a different fundamental use. It is used to understand and utilize sequences by correlating them to observable repeating sequences. The basic time units originated on the observations of the interactions of the Sun, Moon and Earth. To this writer, Minkowski's melding of the Space and Time Concepts seems "an attempt to blend apples and oranges, then call them pears."
He combined a system of measurement, "Space,"

which may be considered as concerning motion along lines. in effect, away from points. to a system of measurement, "Time," which is based on oscillatory or rotary motion. motion that may be considered to have its focus, literally, at some point of origin.
At the subatomic and atomic level, the "OiaS" Model ,produces new tools for studying and correlating interactions between the basic units, the electron-positron and the proton-"conton" sets. This model furnishes at least three useful alternatives to the Proton-Neutron Nucleus Model of the Atom.

The Proton-Neutron Nucleus Model is too ingrained in almost a century of literature to be totally discarded; but should be considered as a bookkeeping device with the "neutron count" being an indication of the portion of the unit which would be considered to be in the "Positron/"conton" form ("Anti-Matter" form) at any given instant. It also may be noted, that in the basic units used in the factor coding of two of the above papers, one may consider that there need be Imbedded "electrons" having their centers of oscillation at the center of each of the basic units.
One "imbedded electron" in the case of the "T^{++}"units and the "D^{+}" units, and two imbedded electrons in the case of the "T^{+}"unit.
Note that since the positive cations of Tritium and

Helium Three could be considered the same units, we could have written "He3$^+$" instead of "T$^+$" or even invented a new--perhaps more correct-- notation, 3^{++}, and 3$^+$ for these cations as components of nuclei.

A count of the "imbedded electrons" that would be found in any given atom by the use of this model, turns out to be exactly the "neutron count."

The work suggests that many of the ideas of molecular chemistry actually apply to the sub-atomic region. Among these ideas are "resonance stabilization" from Organic Chemistry" and the idea of Acid-Base and Redox type reactions being extended to within atoms.

   For example, a possible explanation of the recently reported changing of forms of Hydrogen atoms to Helium 4 units in a process which also appears to change all but one of the other Nickel isotopes to Ni62 may be explained by way of a series of Lewis-Acid-Base -Type reactions between Nickel isotopes and the Hydrogen molecule. The Hydrogen molecule possibly acting first as an "Electron Pair Donor" into to the outer portion of a Nickel atom and then as a "Proton Pair Donor"into the nucleus. (*For any chemist reading this, the last steps would be a cyclic process, Ni62 to Ni64 to Ni66, probably as dication, in each case, with expulsion of a "low-speed-Alpha Particle" and reversion to Ni62.)*

.3. There are enough iconic misconceptions and ideas for extension and application that book-length articles could be probably be written on topics which are barely mentioned here.

4. Another thought:
It might be well, also, for people to check on the not widely known insights of others which have been published primarily on the internet. Among those that come to mind are the work of the Canadian Theorist, Al Zeeper, in his coordination of the various formulations of Gravity, Mechanics and Electromagnetism; and The Cyclic Periodic Chart of William Harrington, which has so many insights that it should be published with a huge instruction book.
Also, there are the thousands of Schemata based on the Periodic Chart that have been produced by Charles William Johnson of Earth/matriX, which offer a wide variety of insights.
The discussions by Jack O'Sullivan published various places on the Internet about "Plasma-Burst Hyper-light" Engines are fascinating, although somewhat difficult to "decode." They indicate that, a half century ago, or more, technology had already far

outrun physics theory, Some of Jack's discussions are included in the book, "Eski's Oscillator/Substance Theory Group, 2008--2011." As of this writing, that book is available in pdf files on Earth/matriX.com. (Free download from the Earth/matriX Forum.) That book could probably be obtained from its compiler, this writer, as a spiral bound copy, autographed even. However, as the total book runs to some three hundred pages, cost for print-on-demand publication and shipping could add up.

If the writer's "Allotted Time" doesn't run out first, a current thought is to revise and amplify an earlier, very private-edition, "Book," called "Cracked Pottery," which could possibly be sub-titled, "Outside the Box and Off the Wall," as it will include not only other material related to this book but, also, essays on subjects having little or nothing to do with physical science and even some putative poems.

This writer finds himself as a "listed contributor" to two Google Books, *Hydrogen* and *Helium,* It is not known what was contributed. It would be nice to know what was thought to be valuable enough to use. Since Google Docs and Google Drive are being used, Google can reprint and use anything they want to. It would be nice if they would tell a person that they were doing so. One might want to include it in the "Resume."

Time to end this. Thanks for reading.

Hasta la vista.

 To contact the writer, Try deanlsinclair@gmail.com or call 1-605-290-2154. *(At the present time, 1000 minutes a month are bought on the cell phone. Usually, six or seven hundred minutes are left over at the end of the month)*   **dls April 12, 2015**